SAY YES TO LED LIGHTING

Our Strategy for Excellence

Jody Cloud

www.YESLEDLighting.com

Say YES to LED Lighting

by Jody Cloud, #1 Bestselling Author

Jody Cloud—Founder + President
YES LED Lighting[R]

Say YES to LED Lighting: *Our Strategy for Excellence*

ISBN-13: *978-1514786741*
ASIN: *B0103LRVYU*

Produced by Rusbarsky Custom Digital Products[R]
www.RUSBARSKY.com

Official Trademark Registration

United States of America

United States Patent and Trademark Office

LED LIGHTING

JODY CLOUD (UNITED STATES INDIVIDUAL)
1521 ALTON ROAD SUITE 536
MIAMI BEACH, FL 33139

Reg. No. 4,812,696

Int. Cl.: 11

FOR: LED (LIGHT EMITTING DIODE) LIGHTING FIXTURES, LED LIGHT BULBS, IN
CLASS 11 (U.S. CLS. 13, 21, 23, 31 AND 34).

TRADEMARK

FIRST USE 12-31-2012; IN COMMERCE 2-1-2013

PRINCIPAL REGISTER

OWNER OF U.S. REG. NO. 4,195,834

NO CLAIM IS MADE TO THE EXCLUSIVE RIGHT TO USE "LED LIGHTING", APART
FROM THE MARK AS SHOWN.

THE MARK CONSISTS OF THE WORD "YES" IN STYLIZED FONT WITH THE "Y" IN THE
SHAPE OF A LIGHTNING BOLT, ALL CONTAINED WITHIN AN OVAL, AND IMMEDI-
ATELY BELOW THAT THE WORDING "LED LIGHTING" IN STYLIZED FONT.

SER. NO. 86-536,972, FILED 2-17-2015

ALICE BENMAMAN, EXAMINING ATTORNEY

Director of the United States
Patent and Trademark Office

iii

YES LED's Professional Associations

Bestselling author and founder of YES LED Lighting, Jody Cloud is officially certified to offer continuing education classes in LED Lighting to members of the following affiliated associations:

The American Institute of Architects

Acknowledgements

I dedicate this book to those whose ideals and beliefs I share and whose desire and direction in life I greatly admire.

The new age of lighting has arrived…and as Jack Welch, Lee Kun-Lee, and many others of their stature will tell you, LED lighting is going to revolutionize the way we light our world.

Thanks to my partners and friends in China who politely answered all my questions, regardless of how strange and dumb they must have seemed over the years.

I appreciate you for letting me into your factories and allowing me to live, work, and even play with your people in their own surroundings. An invaluable portion of what I have learned came from the times spent in your factories. Thanks for giving me free reign to roam the sales floors, the assembly lines, the packing and crating rooms, and everywhere in-between. My experiences were absolutely amazing.

Throughout the course of this book, you will see the name Hamid El-Abd mentioned several times. Hamid is the President of WKK Limited. With annual sales in excess of 800 million dollars, WKK is one the most successful companies in China. Needless to say, Hamid's an absolute expert when it comes to doing business in China.

Hamid has been kind enough to share his amazing keys to success with me, which are the very same keys that he used to propel his own company to the pinnacle of global success. Hamid is a true master when it comes to establishing and maintaining high-level business relationships in Asia.

I am honored to have been taken under his wing and to be privy to his incredible formula for success.

Inspirational Quotes

Jack Welch, CEO General Electric

"If your company does not have a competitive edge, don't compete."

"The team with the best players wins."

"Adapt to the environment at hand."

Lee Kun-Lee, Chairman of Samsung

"To *be* everything, be willing to *change* everything."

Clinton Global Initiative (excerpt)

"In the next few years lighting, as we know it, will change forever, offering us the opportunity to drastically reduce energy use, preventing the release of hundreds of millions of tons of harmful carbon dioxide into our Earth's atmosphere."

About Jody Cloud

Jody Cloud was born in Charleston, South Carolina and was raised in Alexandria, Louisiana. He currently resides and works in both Dallas, Texas and Miami, Florida. His travels take him to China, Hong Kong, and Central and South America, as well as countless cities across America. Jody cares deeply about protecting the environment and improving the planet, a mission that ultimately led to the founding of YES LED Lighting.

Foreword

Hamed Hassan El-Abd *– President, WKK LTD.*

I had the pleasure of first meeting Jody Cloud in 2009 after a very close friend informed me that Jody needed assistance in developing an LED lighting company. Jody's goal was to find high quality and reliable Chinese LED lighting companies to partner with. This goal was quite easy for me to expedite, as my company WKK LTD., (WONG's KONG KING [HOLDINGS], LTD.) is well-known throughout Asia.

WKK LTD., among many other Asian business interests, has been the manufacturer and distributor for the Yamaha LED equipment division for many, many years. As the equipment manufactured by WKK, LTD. has been key in the LED lighting industry for many, many years, I had at my fingertips the knowledge of who were the top manufacturers of LED lighting in China. I also had direct access to the leading executives of these same companies and counted many of them as personal friends. Therefore, it was a simple process for me to introduce Jody to the right contacts in China to establish his company.

I noticed right away that Jody was determined to gain the necessary knowledge that is CRITICAL for doing business in China. Jody crisscrossed China and literally moved into the LED lighting factories in order to learn their processes, strengths, and weaknesses.

My staff and I have worked with Jody over the years and have been impressed with his ability to recognize what it takes to be successful here in Asia. He continues to form amazing partnerships in China and is well-respected by his "Chinese partners" whom have commented to us that he is quite different from other "Americans" who show up in China, looking for a "quick, cheap deal" with "limited" or "no plans" for the future.

Jody is forward thinking and absolutely dialed in as to what it takes to be successful in the LED lighting business. On many occasions, I have heard Jody's name spoken with respect among some of the top LED lighting executives in China.

In closing, I am proud to call Jody my friend, and honored to be part of his success story.

Hamed Hassan El-Abd

Prologue

LED lighting has arrived. The most efficient lighting method known to man is here, and it is sure to revolutionize the world. Consuming up to 90% less power than traditional bulbs, LEDs require only a fraction of the energy used by traditional lighting—while also generating only a fraction of the heat.

Think about this…

If you're lighting a room with traditional bulbs, a significant portion of what you are paying for on your electric bill is countering the heat (via air-conditioning) created by the inefficiency of seriously antiquated lighting technology.

In addition to being vastly more efficient, LED lighting can last much longer—up to 20 years, even when using them for eight hours per day!

Also, LEDs are extremely durable while emitting zero UV emissions and are ecologically efficient. Furthermore, LEDs are also recyclable, and they allow us to reduce our carbon footprint by up to 33%.

People have been using incandescent bulbs for over 100 years now, and it's time for the world to upgrade to a far more sensible and efficient lighting alternative: LEDs.

Given the benefits, it's no wonder why it's so easy for me to get passionate about LED lighting. Frankly, I'm baffled as to why anyone would choose *any* other lighting type over LED.

Yet, most people still do. And there are two main reasons for this.

Reason #1: Price

The first has to do with the cost per bulb. When most people see that the price of an LED bulb is significantly higher than their current lighting, they typically disregard the LED bulb.

But what most people don't realize is that lighting companies continue to make drastic improvements in LED technology, improving the quality and efficiency while at the same time driving down their costs.

According to the U.S. Department of Energy, the cost of LED bulbs has fallen substantially since 2008, thereby making LED lighting much more attractive to consumers. And we will continue to see that number diminish as further advances are made and the supply chain deepens.

Regardless, the price of an LED bulb or fixture is rather inconsequential when you consider the long-term costs of running outdated lighting technology that lasts for a fraction of the time as LEDs.

For example, over the lifespan of an average LED bulb, one can expect to buy dozens of traditional bulbs regardless of the application or wattage.

Furthermore, as you are replacing far fewer bulbs, you greatly reduce your risk of electrical shock or other injury from having to replace your lighting so often.

And the biggest reason why I think the higher cost of bulb is worth paying for is the fact that LEDs are literz 90% more efficient than any other lighting technology on une market.

As a lover of the environment, efficiency is quite important to me.

So in order to have a true and fair cost comparison, one has to take in account more than just the price of a bulb. You have to take a look at the operating cost over time...and the clear and obvious winner here is LED lighting, as it costs a fraction as much to operate as your standard incandescent bulb.

YES	LED	CFL	Incandescent
Lifespan	✓+	✓	✓−
Durability	✓+	✓	✓−
Annual operating cost	✓+	✓	✓−
Light efficiency	✓+	✓	✓−
Environmental hazard	✓+	✓−	✓
Heat emitted	✓+	✓	✓−
Temperature sensitivity	✓+	✓−	✓
Humidity sensitivity	✓+	✓−	✓

As you can see, LED lighting dramatically outperforms traditional forms of lighting in every single category. And now that we've debunked the cost myth, let's take a look at the second reason why people still choose incandescent over LED.

Reason #2: Poor track record

Aside from concerns about expense, the other reason why people continue to buy incandescent bulbs is because the LED lighting industry in general has had a bad reputation over the last few years.

Many LED lighting companies in both the United States and China have gone out of business and subsequently filed for bankruptcy over the last several years. These are tragedies that could have easily been avoided.

The annual failure rate of LED lighting companies in China has averaged more than fourteen plus percent across the industry since 2005.

Major Chinese LED companies such as SINO Light, HIBOLED, and Xu Rui Opto-Electronic Company are among hundreds of Chinese LED lighting companies which have closed their doors over the last 10 years. Samsung recently announced a complete withdrawal from the LED lighting sector, while Philips sold its lighting division to a hedge fund, and Sylvania and others have announced the recent and collective layoff of over 15,000 LED industry-related personnel.

Over the years I have monitored the successes and failures of LED lighting companies across the globe, observing how they've applied these same **10 Keys to Success**—or how they haven't—and, as a result, I have learned a great deal.

My research has shown that most LED companies that fail in the U.S. are directly linked to LED lighting manufacturers that fail in China. While there are multiple reasons for business failures in China, the primary cause is Chinese manufacturers who produce inferior products and are then crushed under the subsequent avalanche of failed products whose warranties they cannot financially support.

Many U.S.-based companies seeking quality LED lighting suppliers fall prey to the belief that all LED lamps are "created equal." Therefore, all too often these inexperienced and uninformed companies sabotage themselves by purchasing these inferior products. The products then fail prematurely and the purchasers discover to their great chagrin that the Chinese manufacturer cannot afford to replace the massive number of failed products.

Eventually, the Chinese manufacturer goes out of business. This scenario ultimately leaves its foreign-based buyers responsible for the defaulted warranties and inevitable lawsuits.

Naturally, the foreign-based LED companies inevitably fail as a result. Therefore, not surprisingly, potential clients frequently ask me, "Jody how can you guarantee that YES LED Lighting will be in business five or more years from now?"

For obvious reasons, this is an important question, and my determination to provide a decisive answer has compelled me to write this book.

I intend to explain the tremendous value of the successful strategy that my staff and partners in the U.S. and Asia have implemented to ensure that YES LED Lighting survives the tests of time.

The story of how YES LED was born is really quite interesting and practically fateful. And in reading my book you will realize just how many advantages we've had since day one.

To be sure, without the *incredible* opportunities and relationships that were basically handed to me, there is no way that YES LED Lighting would be where it is today.

The people I was introduced to ended up being more than great…they were absolutely perfect. This is what allowed me to associate with the best and brightest in the LED Lighting industry.

Over the years, YES LED Lighting has built close relationships with a number of the most highly respected and solvent suppliers and partners in the entire continent of Asia.

 These suppliers and partners cater not only to the market needs of YES LED Lighting, Inc., but also to such industry titans like GE, Sylvania, LG, Toshiba, and others.

When you read about the strategy we have put in place to ensure our long-term success, you will quickly grasp why YES LED Lighting has been able to achieve top ratings in every single LED lighting category.

I will also discuss Hamid's **10 Keys to Success** and how I successfully applied them to create the best brand in the industry.

These are the same **10 Keys to Success** that Hamid used to drive his company's own incredible level of success in Asia.

It is no accident that the YES brand is a trusted leader in the LED lighting industry. YES manufacturers produce products that make us stand out from the crowd, impressing our clients and delivering high quality lighting solutions that continue to impress year after year after year.

While we do recognize that the world has resistance to making the conversion to LED lighting, we also know that in time the technology will speak for itself and people will have the comfort necessary to make the shift.

That's why YES LED Lighting continues to build our reputation for excellence by consistently producing the most reliable products on the market.

As industry leaders like GE, Sylvania, and LG shy away from the LED market, YES LED Lighting will continue to provide the world with a phenomenal lighting alternative.

YES LED Lighting stands out for many reasons.

First of all, I was fortunate enough to be introduced to the winning formula for success, along with the right people to help me get there.

As a result we have also been able build a track record that parallels and/or surpasses the other players in the LED industry.

 In the world of manufacturing LED lighting products, a 3% failure rate is widely accepted during the warranty period. **Poor**

quality suppliers produce products whose initial failure rate can even exceed 25%, often with the remaining balance failing in 3–12 months! This is a frightening, yet all-too-frequent occurrence in the LED lighting industry.

By contrast, every single product produced by YES LED Lighting, Inc. is under the care and guidance of **our** handpicked and rigorously trained quality control team. Our staff members perform rigorous quality control testing throughout every phase of the manufacturing process. The result is that *YES LED Lighting* products experience ***less than one percent*** product failure throughout the warranty period and beyond.

Table of Contents

Chapter 1

History of LED Lighting

In 1879 the lightbulb was invented!

Over 135 years ago!

Yet, when we flip our light switches on, it still doesn't register to most of us just how antiquated our lighting technology is.

However, in recent years, light-emitting diode (LED) technology has seen explosive growth, giving people superior lighting options that…

- Utilize wattage far more efficiently.
- Last up to fifty times longer than traditional incandescent bulbs.
- Possesses a much higher color quality. ← Brightness
- Are available in a wider range of colors.
- Generate significantly less heat.

In years past, LEDs have been used to light things like the tiny bulbs on Christmas trees and alarm clocks all the way up to large and very bright outdoor applications such as digital billboards, street lights and traffic signals.

For years, their potential to light up entire households, office buildings, parking garages, shopping malls, industrial facilities, and virtually every other application which requires lighting was overlooked; but with new advances in the technology, more and more businesses and households are beginning to realize just how efficient LED lighting can be.

Thanks to the emergence of LED technology, gone are the days of inefficient lighting. And because LEDs operate at a much higher Power Factor (meaning that more energy is used for producing light rather than heat), they won't heat up the air in your home or office to anywhere near the degree that an incandescent, fluorescent, or halogen bulb does.

LED technology has come a long way since Henry Joseph Round figured out in 1907 the fundamentals of LED lighting and discovered how to get electroluminescence from a single diode, using only a cat's-whisker detector and silicon carbide.

In case you've never heard of it, a cat's-whisker detector is an antique electronic component that has a very thin wire that touches some type of semiconducting mineral in order to create a point of electrical contact.

This instrument, in combination with the discovery of the phenomenon known as electroluminescence, served as the cornerstone of the technology we see today. It was *this* discovery that allowed scientists to initially conceive LED lighting.

As opposed to incandescent lighting that converts electricity into heat that produces a glow, electroluminescence happens when a material emits light as a result of an electrical current passing through it. It's a much more efficient way primarily because it doesn't require the generating of a large amount of heat just so that it can perform the function of lighting a room.

If you've ever touched a hot incandescent light bulb, you know that the heat produced is not inconsequential. In fact, if it weren't for an inert gas inside of the bulb itself, oxygen would cause the filament to burn up, rendering the light bulb useless. LED technology, on the other hand, has the distinct advantages of energy efficiency and the safety of a cool illuminator.

But little was known about exactly what it was that allowed LEDs to illuminate. In 1927, twenty years after Round's discovery, Oleg Lossev took this new concept a step further and began observing the glowing crystal diodes inside radios while they were stimulated by an electrical current. Despite his efforts, no practical or commercial uses ever came from his research in his lifetime.

In 1955, Rubin Braunstein discovered that certain diodes emit infrared light when connected to a current. And in 1961, Gary Pittman and Bob Biard from Texas Instruments received a patent for infrared LEDs after discovering that a gallium arsenide diode emits infrared light when connected to a current.

However, just one year later, in 1962, the first visible and usable LED light was born! Nick Holonyak Jr, a scientist with General Electric Company, better known as GE, developed the first LED light with a visible red wavelength.

This happened when Holonyak was working alongside another GE scientist, Dr. Robert N. Hall, who was working on forming a semiconductor infrared laser using Gallium arsenide (GaAs). Holonyak changed history by deciding to go a different route altogether by using Gallium arsenide phosphide (GaAsP), which allowed for visibility.

On October 9, 1962, Holonyak became the first person to operate a visible semiconductor alloy laser, and those at GE looked on with amazement at the brilliantly illuminated device. This was such an incredible advance in technology that GE even referred to this new little visible light as "the magic one."

Fifty years later, in 2012, Holonyak when interviewed said he could still remember his thoughts at the time: "I know that I'm just at the front end but I know the result is so powerful...there's no ambiguity about the fact that this technology has a life way beyond what we're seeing." Truer words have never been spoken.

The same year as this major discovery, GE began selling LEDs for $260 each.

Fast forward another decade and you see the first yellow LED, a marvelous creation by a former graduate student of Holonyak named M. George Craford. In addition, Craford increased the red LED's brightness by a factor of ten. These advancements would later prove quite useful in the wide, commercial distribution of LED products.

It wasn't long before Monsanto, the manufacturer of the raw GaAsP semiconductor material, wanted to turn LEDs into a commercially available product instead of a $260-per-bulb investment.

A few years later, Monsanto began large-scale production of alphanumeric displays, which piqued the interest of huge manufacturers such as IBM, Texas Instruments, and HP. These companies began utilizing red LED lights as indicator lights in their products, and LEDs began their high volume commercial production in 1978. Companies started putting very small LED lights in digital watches in the '70s, and by the 1980s they were also in traffic signals and brake lights.

In 1993, Professor Shuji Nakamura took LED to the next level and discovered how to add bright blue to the color spectrum. Once bright blue was available, this opened up the path to the creation of LED's most popular white color. This was such a significant advancement that Shuji Nakamura was awarded a Nobel Prize for Physics for this work.

The landmark year in LED lighting history occurred in 2007 when President Bush signed the Energy Independence and Security Act. This act called for the phasing out of incandescent and certain fluorescent lighting, subsequently sparking a huge boost to the popularity of LED lighting.

More recently, in January of 2014, LED lighting was shown to have completely taken over the energy efficiency world. Now, worldwide brands such as Walmart, Target, Starbucks, and

Marriott utilize LED lighting in their stores and distribution centers due to their efficiency, reduced heat load and lower maintenance costs.

LED lights are capable of emitting bright, clean-looking white light, in addition to multiple different colors, as well. Semiconductor materials can be designed for specific energies, which determines the color that the bulb emits. This technology allows for multi-colored LED lights, and therefore a wider range of commercial products. The most popular colors currently being manufactured are red, green, blue, and crisp white. Your huge flat screen TV uses 3-LED pixels to create wide ranges of color.

Chapter 2

How LED Lighting Works

Light Emitting Diodes have been—and continue to be—a huge advancement in energy efficient lighting. By now, you know that LEDs are very small diodes that are part of an electrical circuit, and are illuminated by the movement of electrons in a semiconductor material. And a diode, simply put, is a semiconductor device that has two terminals that allow the flow of current in one direction only.

Essentially, an LED is an electrical component that emits light when connected to direct current.

First, in order to understand how LED works, it is important to know the different components that the LED is comprised of.

LEDs are made of semiconductors that can be made electron-rich (n-type) or electron-deficient (p-type). Each LED is made of n-type and p-type layers fused together in a tiny crystal. When electric current is passed through the diode, electrons from the n-side combine energetically with the vacancies in the p-side. This combination emits light through a process called electroluminescence.

The color of the light emitted depends on the materials used. This has been an area of intense R&D for decades; for example, gallium arsenide phosphide (GaAsP) is prized for its ability to produce bright red LEDs with high efficiency. The tiny semiconductor crystal, the size of a grain of sand, is bonded tightly to the case to keep it cool; this results in a highly rugged and dependable device compared to the fragile filaments and glass tubes in other lights.

LED lighting is a major advancement in energy efficiency, durability, and reliability, all due to their material and construction. LEDs have very long life because they have no metal fatigue, filaments, or oxidation, and they don't shatter because there is no glass to break.

They should be viewed entirely separate from incandescent and fluorescent lighting because the way they operate is vastly different. An LED light with a 100,000 hour life expectancy can last up to 91 years. Imagine only having to change your light bulbs every 10-15 years! This may sound impossible, but this is exactly where this technology is headed.

By understanding just a little bit of the science of the design of LEDs, one can clearly see why they are a superior lighting option. LEDs contain a plastic body that is encased around leads and a semiconductor, and the light shines through the plastic body. Most LED bulbs have a rounded shape to focus the light in one direction. LED lights are known for a high level of brightness and, at the same time, are highly efficient and reliable. An LED is a different kind of light entirely, and does not produce potentially damaging and harmful UV light.

Let's now take a closer look at the actual design and build of a standard LED bulb. Looking at the diagram below, the first thing many people notice immediately is how similar it looks to a traditional light bulb. In some ways this is simply clever marketing, but since LED bulbs are designed to replace traditional light bulbs, they have to be able to fit and plug into the existing fixtures (the same design technique applies to all LED bulbs and fixtures relative to how the components function).

'Exploded' LED Diagram

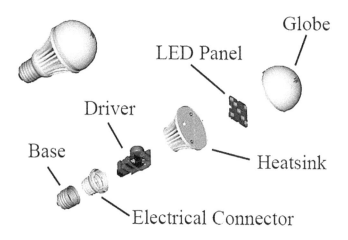

Let's talk briefly about each part, starting with the base and moving up (this same dynamic applies to every LED bulb and/or fixture regardless of application):

- **Base** – Comparable to base of traditional light bulb. Designed to be plugged into existing sockets.

- **Electrical Connector** – Connects to the flat connector at the bottom of the Base and transfers the power up to the Driver.

- **Driver** – Converts the AC power of a structure into the DC that LED bulbs need in order to operate.

- **Heatsink** – Aluminum fins disperse heat away from the electrical components, helping preserve the life of the bulb.

- **LED Panel** – A printed circuit board that houses the LEDs (usually around five or ten for most standard applications).

- **Globe/Lens** – Placed on top of the LEDs to help disperse the brightness and provide a nice, even glow of crisp, clean white light.

The LED light has an all-around simple, strong, and versatile design. Since light emitting diodes rely on the movement of electrons creating light, they don't run into issues such as burnouts, because electroluminescence is not a process that eventually wears out. This makes for a highly reliable bulb that can last you for years.

For example, suppose you were to use an LED light for 12 hours per day, every single day. Given that LEDs have a 50,000-hour lifespan, you wouldn't have to change your light bulb for more than 11 years!

Electroluminescence also allows for a quicker process of the light being turned on, therefore it's the perfect solution for traffic lights and signals. LED brake lights turn on virtually instantaneously, while incandescent bulbs take more than a tenth of a second. At highway speeds, LEDs provide ten extra feet of safety!

Another reason why they're popular in outdoor fixtures, such as traffic signals and parking lot lighting is that LEDs last much longer than traditional lighting which are very expensive and time consuming to replace.

And if higher efficiency, brighter, longer lasting, and more versatile lighting isn't enough, LED lights are also extremely safe and eco-friendly. They contain no mercury, no hazardous glass, and are significantly better for our environment.

In fact, by using LED lights, you can decrease your carbon footprint by up to one-third! And, since LEDs contain no hazardous materials, there is no need to be concerned about disposal.

In addition, if an LED happens to break in your home or office building, you don't have to worry about any toxic chemicals such as mercury.

Chapter 3

The Box

If someone had told me that I would one day be traveling to China for the purpose of creating a global LED lighting company I would have thought they were insane. Little did I suspect what the future held for me, as soon I'd be making routine trips halfway around the world to China. Now, several years and over 500,000 air-borne miles later, I've come to call China my second home. I've become just as comfortable on the streets of Chinese villages and cities as I am in the U.S. (I do not speak Chinese, although I am a very slow and steady work-in-progress).

Yet, as fate would have it, in August of 2007, Herbert Miller, a close friend of mine from Asia, gave me a gift that would forever change the course of my life.

This gift was a simple cardboard box full of strange-looking glass and aluminum LED light bulbs, that if not for a random lighting storm would have been destined to languish in my closet.

The force of the lightning generated by this violent storm was so fierce that it created a power surge, blowing out a number of my existing halogen light bulbs. I was left with the choice of being left in the dark, traveling to a hardware store in really bad weather, or using these really strange-looking light bulbs.

I made the simple choice to install the LED lamps and when I flipped the switch and the LED lamps powered on, I was immediately in awe. The LED lights filled my kitchen and living room with the most amazing quality of light I had ever seen. The light was clean, clear, and crisp, without being over-the-top. And I would later learn the reason why: the CRI, or color rendering index (essentially the light clarity), of LEDs is vastly superior to traditional lighting.

The next thing I noticed was how cool the air in my home had become. That's when I realized that the newly installed LED lighting was generating significantly less heat than the twenty-two, 400+ degree halogen bulbs that had previously resided in my kitchen and living room.

By morning, I had installed every single LED lamp I received from Mr. Miller.

It quickly occurred to me that I had stumbled onto a spectacular new technology. At this time, (now seven plus years ago) LED lighting was virtually unheard of; however, it did not take a rocket scientist to see that LED technology was poised to one day explode onto the world stage. The opportunities for tremendous energy savings, vastly improved visual environments, and major reductions in greenhouse gases, were overwhelming. When combined, these attributes offered a huge "win" for everyone; and as an entrepreneur, I cannot begin to tell you how excited I was when I initially realized the obvious potential for LED lighting.

I decided to arrange a meeting with a number of colleagues who had built successful businesses both domestically and internationally to discuss this incredible technology and its potential.

They too were immediately blown away when I demonstrated how the LED lighting worked in my home. We subsequently brainstormed (over a couple of beers of course) about the obvious long-term impact and huge market for LED technology. The ensuing conversation left me absolutely convinced that there was an incredible business opportunity right in front of me.

I had undoubtedly found myself on the path of one of the greatest technological advances in generations. Consider this: LED lighting offers a massive reduction in energy use, as well as drastically reducing the tremendous amount of heat generated by conventional lighting…not to mention the fact that its applications are practically **infinite**.

As my goal was to form an LED lighting company it was crucial to I learn every facet of the industry, including the components involved the manufacturing process and, most importantly, the marketplace.

My research quickly revealed that over 90% of the world's lighting is manufactured in China, therefore, that's was where my path was obviously going to take me.

But how in the world does one even go about learning how to do business in China? The answer? When in doubt—**Google**!

One of the first things I did was Google topics such as "doing business in China," "the pitfalls of doing business in China," "what to be aware of when doing business in China," and dozens of similar queries. I then reached out to a number of friends and colleagues who had personally done business there.

Their common advice was to **be very careful!**

Needless to say, I was prepared to face a lot of obstacles. The greatest initial challenge was overcoming the language barrier. Chinese is such a unique language with many dialects, (not to mention 10,000+ characters, not letters like in English) and there are no direct cross-translations from English to Chinese or vice-versa.

But, even still, my greatest concern was how to find the right suppliers. I had heard first-hand and read online the horror stories of people and companies who were totally victimized in faulty business dealings involving Chinese companies.

China is, after all, over 10,000 miles away, on the opposite side of the world, safe and secure from most laws behind the borders of a communist country.

However, I knew that in order to be in the LED lighting business I had to travel to China and pay whatever price necessary to systematically identify and partner with high quality, reputable, and reliable Asian manufacturers, just like other successful entrepreneurs had done in the past.

This process, by the way, has (so far) accounted for over 600 hours of actual travel time, 500,000 miles of air travel, and nearly 7 years of my life. *Sounds crazy, right?* Well, let me tell you: the hours, days, weeks, and now years I've spent meeting suppliers, conversing with experienced LED lighting engineers, participating in the design of newer generation products, working on factory assembly lines, and being immersed in every segment of the LED lighting industry has given YES

LED Lighting a clear and definite advantage. It has been well worth every mile flown and every minute spent working and conversing with the owners, engineers, and laborers within the factories with whom YES LED Lighting has partnered over the years.

Chapter 4

Finding a Major Ally

To this day, the single most important move that I ever made was to reach out to my friend, Herbert Miller, who gave me those first LED bulbs as a gift.

As a high-profile executive, Herbert had spent several years living and traveling in Asia both before and after his retirement. I was hopeful he might have a contact in China to whom I could turn for much-needed advice and guidance. And, as fate would have had it, he did. Herbert soon introduced me to one of his closest friends.

However, at first Herbert, forever the comedian, would not tell me *who* his colleague was, only that he had the "perfect guy" to help me. He really had me going for a while, but when he finally divulged who his "friend" was, my jaw hit the floor!

He ended up introducing me to Hamid Hassan El-Abd, who is the President of WKK Limited. As the renowned, driving force behind one of the most successful companies in Asia, Hamid's accomplishments and roadmap to success are legendary in Asia.

WKK, a manufacturing giant in China, was founded in the 1950s and employs a workforce of over 20,000. Among other partnerships, WKK is the exclusive manufacturer and distributor for Yamaha's LED Equipment Division in Asia.

Imagine if you were a kid showing up at the park for the first time to play basketball and your coach was Michael Jordan.

A million words would never be enough to describe my excitement. No amount of money could have purchased the invaluable information and introductions that Hamid and his staff has shared with me over the years. Not a day goes by that I do not give thanks for his help.

After being introduced to Hamid El-Abd, and being made aware of his proven **10 Keys to Success**, I truly felt a sense of divine intervention compelling me to initiate my journey to China and into the LED lighting arena.

I am certain that if not for my chance opportunity in meeting Hamid, being made privy to his keys to success and being granted access to the invaluable resources and relationships at his disposal, that YES LED Lighting and I would have fallen into the same traps that have led to the demise of hundreds, if not thousands, of companies involved in LED lighting manufacturing in China.

Chapter 5

Preflight

So here I was, determined to travel to the opposite side of the world.

There were several important questions to be answered.

How do I get there? Are there any special travel documents or visas required? What airline(s) do I use? How long is the flight? Can I fly direct? What is the weather going to be like? Is the Chinese language a big barrier to communication or are they mostly bilingual?

Traveling to China does indeed require a visa. And since I've never been one to procrastinate, I immediately applied for a Chinese visa, for which a great deal of paperwork was required. Passports must be presented to a Chinese consulate for the visa to be issued, and all travel itineraries need to be provided.

The visa cost $250.00 and the options were for a single or multiple entry, good for one year from date of issue. For flexibility I chose the multiple entry option and within 30 days my visa was issued, attached to my passport and returned to me by the Chinese consulate in Houston, Texas. I quickly learned that there are many flights to China and Hong Kong, but that there were NO direct flights to either of these destinations from Miami, which meant that I was going to have at least one connection.

For my flight to Asia I chose American Airlines. Since I have been an executive level customer for many years, I was entitled to significant benefits including the Admirals Club, free checked bags, double miles flown and, most importantly, cabin upgrades.

The route I chose was Miami to Dallas (2.45 hours), Dallas to Tokyo (14.5 hours), and ultimately Tokyo to Hong Kong (5 hours) with a change of carriers to Japan Airlines. I must mention that there is also delay time between flights that collectively add around five to eight additional hours each way. All in all, this round trip flight would take more than seventy hours of actual flying and layover time. And let's not forget about the twelve hour time difference.

When it comes to jet lag, you simply have no idea!

For obvious reasons, I chose to plan my trip to coincide with one of the two largest LED Lighting Trade Shows/Conventions in China. The LED Spring Lighting Fair takes place every year in Hong Kong, late February through early March, while the Fall Fair is in October.

It was crucial that I secured the services of a capable, reliable, and industry-experienced interpreter. In my situation, I especially needed an interpreter who had experience in the LED Lighting industry. Fortunately, Hamid had the foresight to obtain the best possible interpreter, a Chinese man named Peter who was an electrical engineer and fluent in English, with a long history in LED lighting technology and its evolution!

Peter would ultimately spend several days with me, a novice, explaining the technology, history, and future of LED lighting.

The stage was set and I was seriously pumped up. Now it was simply a matter of counting down the days to departure.

Chapter 6

Destination: China

My boarding time was 8:00 am, so I was careful to arrive at the airport by 6:00 am to make sure everything was in order for my check in.

Keep in mind that almost everything up to this point had been a new experience and I couldn't help but wonder what I was getting myself into.

However, despite being somewhat nervous about traveling to the "other" side of the world into a virtually unknown environment I reminded myself of the significance of this journey and my determination to bring my vision to life.

With my passport and visa in hand, I walked up to the ticket counter to pick up my boarding pass.

My first flight to Dallas left on time and once I arrived, I had plenty of time make my way to the connecting American flight departing to Narita Airport, located in Tokyo, Japan.

The flight from Dallas to Tokyo was a back-breaking 14.5 hours, and the first thing I thought after clearing customs in Tokyo was that I wanted something to eat that was **not** airplane food.

So I casually strolled by restaurant after restaurant that served food that I've never eaten (or even seen) before. Being unaccustomed to authentic Asian food, I opted to have my first Big Mac and fries in over ten years. (I never thought I would hear myself as an adult say "THANK GOD for McDonald's!")

Three hours later, I was more than ready to drag my exhausted self aboard my last flight. My JAL flight departed Tokyo on time and I was so tired that, frankly, I could not keep my eyes open and ended up sleeping for over three of the five hours that we were in the air, all while sitting in one of the most uncomfortable, rock hard seats imaginable.

I did stir for a few moments during the food service. However, I took one look (and sniff) at the fish parts and rice the flight attendant offered and decided to pass. Thank goodness for protein bars I had stashed in my backpack.

Five hours later we touched down in Hong Kong and my adventure of a lifetime was about to begin.

Chapter 7

Hong Kong

"Manhattan on steroids" is the best way to describe Hong Kong. For those unfamiliar, Hong Kong is composed of a chain of islands and is a Special Administrative Region of the People's Republic of China, flanked by the South China Sea and the Pearl River Delta.

Hong Kong has its own currency known as the Hong Kong Dollar, while mainland China's currency is known as the Yuen or RMB.

The Hong Kong International Airport is located in an area called the "New Territory," about 45 minutes from the central business areas of Hong Kong and its sister islands. Massive bridges and freeways connect them all. The taxi fee to the HK Central Business District is about $35.00 USD.

The architecture and the incredible size of the buildings are absolutely spectacular and if I had not seen it with my own eyes, I would never have believed it.

Just imagine a mountainous Manhattan surrounded by other mountainous islands…times ten! Anyone possessing the proper documents can travel to mainland China from Hong Kong by various means, including train, boat, bus, taxi, private car, shuttle van, water ferry and even helicopter for the wealthy.

Just be certain to have your Chinese travel visa and your

passport in-hand and be ready for inspection and verification as the immigration and customs lines are extremely long but move very quickly.

It is important to mention that Hong Kong is the "financial heart" of Asia and the amount of money flowing through there is staggering. Virtually every multi-national company has an official and very expensive presence there.

Hotels in Hong Kong are also expensive. Hamid recommended the Holiday Inn Hotel at the Golden Mile (the epicenter of Hong Kong's business district). At $350.00 USD per night, it was a bargain and was conveniently located close to his offices as well as the Hong Kong Convention Center, where the Spring LED Lighting Fair would take place.

The cost of real estate in Hong Kong (called HK by international travelers) is startling and can make real estate prices in Manhattan or Los Angeles seem like a bargain by comparison.

Chapter 8

Meeting Hamid El-Abd

I was fortunate to have my new friend Hamid as an ally. It was extremely serendipitous that WKK Limited was the exclusive manufacturer and distributor of the Yamaha SMD machine, a staple and critical component in LED lighting manufacturing. These computerized machines embed the LED chips on the circuit boards. [Next, the boards are connected to the power supplies, which are then inserted and sealed into the housings that hold the LED components.]

My first meeting with Hamid was scheduled for the evening of my arrival. Hamid, aware that I was unaccustomed to Asian food, mercifully made reservations at a spectacular Italian restaurant, and this bustling restaurant was completely packed. Though I was exhausted to the point of being a little disoriented, it was time to get down to business.

Hamid and I had exchanged dozens and dozens of emails and Skype calls during the months prior to my arrival in Hong Kong.

The specific purpose of our numerous correspondences had been to identify my goals and to create an outline of how to create a successful company relative to business relations in China.

Our dinner meeting served as the culmination of our discussions regarding understanding and implementation of the

10 Keys to Success. Not to mention the fact that I was finally able meet an industry icon and thank him in person for basically handing me the roadmap to incredible success.

Over the last several years, I have met scores of people who have experienced tremendous disaster and heartache in their own business dealings in China. As an observer, I have witnessed firsthand the carelessness with which many businesses and individuals—including the representatives of large corporations—conduct themselves in China.

As dinner progressed and Hamid and I discussed at great length the failures of many large companies, we invariably identified the lack of attention to one or more of the **10 Keys to Success**.

To be sure, I reference these **10 Keys** every day not only to avoid disaster in business, but also to continue realizing ongoing success!

These **10 Keys to Success** also serve as a constant reminder that the "shortcuts" are commonly the path to disaster and ruin.

The reality is that these **10 Keys** must be followed in their entirety for every transaction.

If just one is ignored your business is going to take a major, major hit sooner or later.

Chapter 9

10 Keys to Success

These following **10 Keys** became my mantra for **Success**. The order of these is not specifically important, but the importance of following each of them is critical.

1. Locate the right factory.
2. Meet the chairman.
3. Tour and observe the factory components.
4. Identify the factory's strengths and weaknesses.
5. Identify the factory's current customers.
6. Assess the factory's quality of products.
7. Assess the factory's consistency.
8. Assess the factory's ability to communicate.
9. Assess the factory's responsibility.
10. Assess the factory's accountability.

I can recite these points in my sleep, and I repeat them to myself as a literal checklist when meeting with potential and ongoing suppliers and partners in China.

The **10 Keys to Success** are easy to grasp and are all based in common sense.

Let's analyze and discuss them one by one.

Key 1—Locate the right factory.

Normally this is achieved by utilizing extensive internet queries, which are supplemented by attending trade fairs and then following up with factory visits.

This process sounds easy, but in reality is very risky as the buyer has absolutely no way to be sure that the suppliers chosen are reliable. That is, it is virtually impossible to know how solvent and stable these companies are financially, or whether they produce consistently high-quality products. One is basically "shooting in the dark" by choosing suppliers this way.

There is also the ever-present challenge of ensuring that the companies you encounter online or at trade shows are even the actual factory.

To be clear, there are far more trading companies to be found online and at trade shows than actual factories. And, believe me, these trading companies are typically far more sophisticated with their marketing and advertising abilities and materials than the factories themselves, which leads many buyers to the false conclusion that they are doing business with an actual factory. Therefore, while the task is nowhere as easy as it may seem, it is absolutely crucial to locate, identify and subsequently meet with the actual factory.

Thankfully, this initial first step was simplified for me as Hamid and WKK personally introduced me directly to the owners and operators of several of the oldest, largest and most stable LED factories in Asia. This is how YES LED Lighting was able to "hit the ball out of the park" from day one.

Key 2—Meet the chairman.

A Chinese business owner is typically known as the *chairman*. It is critical to meet him and his second-in-command, the *vice chairman*. The reason for this is very simple. By taking the time to meet the chairman and vice chairman, and to stay in contact with them, one is able build a personal relationship. The more effort taken to strengthen this relationship, the more valuable the relationship is. This is a very different relationship than if one simply creates and maintains a relationship with the sales staff who have absolutely no decision-making ability and who really only want you to place an order, and will often tell you anything to get that order from you.

It is very important to pace yourself while conversing with the owners of these companies. The reasoning, Hamid made clear to me, was that you must not over-promise or over-commit. The resulting loss of credibility from "under-delivering" is swift and permanent. There is absolutely nothing to be gained by promising massive orders or asserting that you will soon become their biggest customer, unless, of course, you really are there to place a major order and kick off a large-scale partnership.

Believe me, they hear ridiculous promises from foreign visitors every day that never even come close to being delivered.

The factory owners will meet with you to hear what you have to say. My experience has been that they really enjoy holding these meetings and I have found many of these industry leaders to be interesting, engaging and, in some cases, extremely entertaining. Nevertheless, always remember that credibility is paramount with them. Therefore, be careful with your words.

To put it bluntly, do not ever promise what you are not prepared to deliver.

I cannot begin to stress the value I have gained by initiating and nurturing these relationships. They have truly proven to be priceless in conducting business in China.

In my case, due to the high-level introductions that I was afforded, every single factory owner took us on a personal tour of the factory as well as taking us to a lavish lunch or dinner. And, believe me, once the workers and sales people see you touring the factory with the chairman, the level of respect you receive changes exponentially!

Key 3—Tour and observe the factory components.

This is always an interesting visual experience for me. It tells volumes about the attitude and pride of the workers, as well as the smoothness and organization of the internal processes. The presence of organization and pride are critical stepping-stones on the path to success at every level.

The things I was advised to assess on a factory tour:

- Is the factory organized? Is it clean?
- Are the workers polite?
- How are the workers dressed? Are they wearing uniforms?
- Do they take pride in their appearance at work?
- Are the component storage and stocking areas organized?

- Do the factory managers, who typically speak English, effectively and factually, communicate necessary and critical information?
- Are the forklifts and other vehicles used to load and deliver products in good condition?

If the answer to any of the above is no then you are likely in the wrong place and need to move on as politely and expeditiously as possible. It will be only a matter of time before their lack of organization and flawed internal procedures and bad attitude become *your* problem.

Key 4—Identify the factory's strengths and weaknesses.

It is important to identify the factory's strengths as well its weaknesses.

Each factory always provides **top** quality products within **one** lighting category. It is critical that this product category is identified and targeted for manufacturing.

This is obviously a very important strength to recognize.

YES LED Lighting manufactures the best LED lighting in each category by partnering with the best factories in those specific categories.

For example, a commonly made mistake is manufacturing street lighting at an LED tube lighting factory, or manufacturing tube lighting at a floodlight factory, etc. (the point here is to ONLY utilize a factory relative to what its KEY product is).

A factory might have the ability to manufacture great products, but can be totally inept at quality control or meeting delivery schedules, for example.

Quality of communication is another important element to be aware of when conducting business in Asia. Factories whose employees have poor communication skills can result in disasters on multiple fronts.

As one can see, there are many factors that are crucial to successful international business transactions. Therefore, it is vital to identify any potential weaknesses in areas such as communications, quality control, delivery schedules, shipping requirements, etc.

These and other important areas of concern are generally easy to identify and address as long as one knows what to look for, as they become very obvious, very quickly.

Refusal to recognize significant shortcomings for what they are can yield disastrous results. It is simply common sense to be strategic regarding the positive *and* negative effects of supplier-partners' strengths and weaknesses.

I can state definitively that many of our competitors have taken the easy road of convenience while unknowingly sacrificing quality of products in various lighting categories. In addition, major problems can and will arise if the supplier has problems in areas such as arranging and facilitating, containing, loading, and subsequent shipping.

Key 5—Identify the factory's current customers.

Before my very first pre-arranged factory visit, Hamid gave me a cryptic tip to check out the shipping area. He said that there was important information awaiting me there and he told me to call him when I arrived in that area.

As requested, I called Hamid when I reached the shipping area. His first words were, "What do you see?"

I responded that I saw a huge warehouse area with dozens of people putting products in boxes and then packing them into cartons, which were then placed onto pallets for shipping.

He repeated, "This is very important: what do you *see*?"

Still not understanding his meaning, I started to repeat what I had previously stated.

He politely interrupted me. Right about then I began to feel a little like we were in a scene from the movie "Karate Kid" with Mr. Miyagi saying again, "What do you *see*?"

Hamid then requested of me, "Read off the company names on the boxes you see."

Aha! I immediately got his point as I began to read off the names of some of the most trusted lighting companies in the world: "Philips…Lowe's…MaxLite...etc."

Hamid asked me if I knew who those companies were.

Of course I did!

He then said, "If these industry titans can trust their business and reputation to this factory, don't you feel like you can do the same?"

I responded, "I thought Philips, LG and these other companies had their own factories."

I will never forget his answer, which was, "Why should Philip, LG and those other companies build and maintain their own factories when they can use the existing Chinese factories?"

What I learned in this moment was that I had access to the exact same factories as the major international lighting companies such as Philips, GE, MaxLite, and others. I almost fell over (and not just from exhaustion).

Key 6—Assess the factory's quality of products.

With Chinese products, it is crucial to understand that not all products that look the same actually work the same. Nearly every product made in China can be (and often is) copied on some level. Imitations and copies mimic the component materials, packaging and even the authentic-looking cartons with bogus serial numbers that appear to be from companies such as Texas Instruments, Intel, and others.

Therefore quality control and component verification and authentication is critical. In order for YES LED Lighting to be on the strongest possible footing with quality, consistency, and authenticity, we employ our own quality control team. We make sure that every single product is correctly tested based on our own rigorous guidelines. This process yields phenomenal results as we frequently catch and therefore prevent the shipping of products that do not meet our set of high standards.

This process also serves to provide our Chinese partners with a clear and respectful understanding that we know what we are doing as far as quality control. It is a foolish and common practice for inexperienced and poorly advised companies to defer quality control to the Chinese factories in which the products are made. The lack of understanding of how to establish and maintain strict quality control is a major reason why a number of our competitors have been forced to close their doors and many others are doomed to do so in the future.

Key 7—Assess the factory's consistency.

While consistency is often an issue that can be resolved with quality control, it is important to note that consistency is a huge problem in the LED lighting arena.

Consumers in the United States tend to take for granted that certain products are from reliable sources and that the product quality and functionality will likely be the same every time. For example, we can reasonably expect that a size 9½ shoe is always a size 9½ (and for both shoes).

We are very careful in making sure that YES LED Lighting products are consistently manufactured to the exact same specifications and standards.

If you do not carefully scrutinize your suppliers and put specific safeguards in place to ensure consistency, then you should not be surprised if identical but separate orders yield different levels of appearance and functionality.

I have heard horror stories relative to this issue. Foreign companies receive samples from China and then place a large production order, only to receive products that are not even close to the samples that they originally received. Our strict quality control procedures eliminate this possibility.

Thankfully, I was advised early on to be cautious of this. I had been fully versed and therefore know what to look for. As a result, YES LED Lighting always meets the highest standards in quality, functionality, and consistency.

Key 8—Assess the factory's ability to communicate.

The phrase "communication is king" is applicable to any environment. Therefore, common sense tells us that the ability to communicate clearly is especially crucial when doing business half of a world away in China.

It is very important to have a contact within the facility that speaks and writes in English, as we insist that every production contract be written in English.

Additionally, I have found it essential to maintain at least two on-staff interpreters who can speak and write in both English and Chinese.

It is also extremely important to have a clear and open line of communication about any problems or challenges that may arise during manufacturing which may affect shipping and U.S. delivery schedules. Fortunately, I was strongly advised as to the value of communication in creating and maintaining a successful international business.

Key 9—Assess the factory's responsibility.

The necessity of responsibility is obvious in any country and any culture. However, we have all learned that responsibility is essential—but not exactly prevalent—in today's business environment.

We can all relate to instances and events where no one wants to take responsibility. The same holds true with international transactions, whether these transactions are service-related or are product-based. The last thing you want to experience is a product breakdown on a massive scale, only to find that the manufacturer won't accept responsibility or assist you in resolving problems.

There are scores of factories and trading companies in China who work at arm's length via the Internet and/or telephone who will leave you hung out to dry should you have a product problem or require technical assistance.

The stories I have heard and the catastrophes that I have personally witnessed are heartbreaking, unfortunate, and mostly preventable. Again, without the necessary due diligence in choosing foreign partners, a company can leave itself exposed.

Key 10—Assess the factory's accountability.

Some believe that responsibility and accountability mean the same thing. I can assure you that they certainly *do not*, and instead hold quite different meanings.

Responsibility means, "Yes, I am the one who crashed into your car."

Accountability means, "I am going to repair your car."

Therefore, it is prudent to ensure that goods shipped from 10,000 miles away are provided by companies that will hold themselves accountable for product failures and other challenges that can arise.

As stated earlier we are very careful and methodical to ensure that every order and every shipment is based on a written and executed contract that defines the exact terms of that specific transaction.

The contract must include the exact specifications, components, and functions of the goods produced. Furthermore, it must outline the specific responsibilities and accountability of each party during every step from date of purchase through the end of the product warranty period. No detail is too minor.

It's essential to have suppliers and partners who honor their warranties and hold themselves accountable for failed products by replacing them or providing credits against future orders. Yet, hundreds of companies, including many U.S.-based LED lighting companies, continue to fail due to their relationships with suppliers and partners who do not hold themselves accountable.

Do not make any assumptions. Know your supplier-partner and get everything in writing!

Chapter 10
The Hong Kong LED Lighting Fair

There are two major LED Lighting Trade Shows every year in Hong Kong, known as "Lighting Fairs." Four days in duration, these trade shows are each attended by more than 10,000 exhibitors as well as 500,000 attendees seeking to purchase LED lighting. The exhibitors include all sizes of Chinese companies, from very small to global enterprises.

What many people don't realize is that a large number of exhibitors are actually Hong Kong-based companies that should not to be confused with an actual factory. Commonly known in Asia as trading companies, these companies purchase products from the actual factories in China and then sell them under their own company name. This places the buyer at serious risk because the selling company is not the same as the actual factory and the buyer's relationship is not with the factory but with an independent third party.

It is important to mention that in these situations the warranties are not with the factory, but rather with the trading company. This means that if the trading company goes out of business, your warranty dissolves along with it, regardless of whether the actual factory is in business or not.

I was extremely fortunate to attend the Lighting Fair with Peter, the engineer from Hamid's company, WKK LTD. Not only was Peter an expert in the area of LED lighting and its technologies, but he also spoke flawless Chinese and English. It was nothing short of a miracle to have been accompanied by an incredible LED technician and an amazing interpreter rolled into one.

As we walked the rows of the trade show, Peter would point out various products, providing me a tutorial of their construction and function. He showed me what was good about their design as well as what was flawed.

He keenly weeded out the trading companies, quickly dismissing them altogether. I marveled at his skill, as many Hong Kong trading companies are masterful at convincing you that they are really the actual factory.

However, I am now pleased to say that I too can see through their illusory veils.

Since over 75 percent of the exhibitors at the Hong Kong Lighting Fair ended up being trading companies, the time it took us to walk the exhibitor floors was greatly reduced. We still spent two full days touring the eight plus miles of aisles with booths on either side.

Today, with more than a dozen LED Lighting Fairs under my belt, I am still confounded by the number of people from the U.S. that I see crammed into the booths of the Hong Kong trading companies and placing orders at the LED Lighting Fairs.

Occasionally, I pause at the edge of their booths, listen to their conversations and watch them place large orders, believing that they are dealing directly with a Chinese factory. They never take the time to visit and verify these "factories," to their detriment.

At any rate, my inaugural LED Lighting Fair turned out to be a smashing success. I was able to observe the LED industry up close and personal and see LED technology on display while receiving a powerful industry tutorial.

We were also able to schedule meetings with several suppliers in China to complement the number of pre-arranged meetings Hamid and Peter had already put in place.

Chapter 11

Choosing Factories

Regardless of the industry or product, there are few words to describe what it is like to move about in a country that is industrialized and populated to the degree that China is. From town to city, from one province to the next, the sheer volume of production and transportation is staggering.

The center stage for the world's manufacturing is China. The remote rural villages from which millions of Chinese workers migrate, the factories that manufacturer the goods, the trucks that move the goods, the containers that hold the goods for transport, the warehouses that hold the goods and containers, the massive container ships and, in many countries, the actual ports themselves all play a role in making China the world's source for goods.

China has created an unimaginable vertical market that touches every segment of our world. It is no coincidence that China continues to invest in virtually every nation in the world.

In fact, China owns over $1 trillion in U.S. Treasuries, which amounts to over 21 percent of the U.S. debt held overseas and about 7.2 percent of America's total debt.

As my goal was to source the highest quality products in the marketplace, my main focus was to only visit the factories that produced the highest quality products.

I visited some factories with Hamid's staff, but he additionally recommended that I schedule a few factory visits myself.

It is important to note that the majority of the companies who wish to find suppliers in China use what Hamid called the "Western" way, which means relying on word-of-mouth, the internet and/or the LED trade fairs.

In my case, I used the internet and scheduled visits with a number of trade fair vendors who seemed to have quality products in order to schedule my factory visits (Western-style).

However, the vast majorities of the suppliers and companies that I found through the "Western" method were not even Chinese companies. Instead, these companies were the Hong Kong trading companies mentioned earlier who buy goods in China and resell them; wherein the warranty is not with the manufacturer, but with a trading company instead. This creates a major challenge when problems ultimately arise.

Visiting factories the "Western" way was the absolute OPPOSITE of the factory visits arranged through Hamid and his incredible staff.

I observed many poorly organized, horribly maintained facilities, as well as employees dressed in shabby and unprofessional attire, shuttling me around in old and unreliable vehicles. Their drivers frequently chain-smoked in the car on the way to and from the factory while often screaming on a cell phone for the 90-minute drive to and from the factories.

Many of these factories had floors filled with people working in contaminated and completely disheveled environments. Dogs and children were commonly seen walking, playing, and lying down in what should be "clean areas." At other times, I observed people sitting on filthy floors assembling products, chain-smoking cigarettes and eating while they worked.

I noticed that components and semi-assembled products scattered in complete disarray was actually the norm for these places.

To be sure, I was shocked at the total and complete confusion and negligence of the workers and their managers, as well as the disregard for their work and the products being produced.

Hamid had explained to me early on that a huge number of Chinese factories fail, going completely out of business, leaving their customers to their own devices (of course, this dynamic does not ONLY apply to the LED industry).

After visiting some of these factories, it was readily apparent the types of things that lead to such a widespread failure rate that is rampant in the LED industry.

The continuing failure of these companies and the poor-quality products they produce have had a tremendous negative impact around the world, and have given the LED lighting industry and Chinese suppliers a serious black eye. The overall impression of product quality, consistency, and actual product warranty suffers and Chinese manufactured LEDs as a whole are placed under a cloud of doubt.

It's important to remember that China is on the opposite side of the globe. Correcting problems once products reach their foreign destinations is costly and, frankly, virtually impossible.

In addition, the distance creates a logistical nightmare for U.S. companies when these same faulty products are subsequently shipped from a U.S. company to hundreds of consumers in hundreds of different locations.

This is just one of many reasons why it is crucial to adhere to the details listed in our **10 Keys to Success**.

These **Keys** have served me well in avoiding the typical, unnecessary and costly errors made by many of our would-be competitors.

The 10 Keys to Success Method

Having toured factories using Hamid's resources, with our **10 Keys to Success** as a constant reference, and with Peter at my side, I encountered nothing other than state-of-the-art factories using the most modern and sophisticated equipment and technologies in the world. Their employees were multi-lingual people who dressed and conducted themselves as polished, well-trained, and talented professionals.

The "WOW" factor was palpable!

As advised by **Key 2**, we met the chairmen and directors of each of these gigantic facilities. These were places with the size and prestige of BMW or Volkswagen plants.

The factories that Hamid guided me to were all expert in different key product categories, as discussed in **Key 4**. They all had their own strengths and weaknesses when it came to different categories.

For example, the top LED street lighting manufacturer would not be the top LED tube manufacturer and vice-versa. And since I sought the best LED lamps from the best LED lighting suppliers in each lighting category, I needed to know where to look and what to look for.

This is one of the most important things I learned, and my understanding of this has enabled me to construct a company that offers the best quality products in every lighting category.

I continue to see U.S. companies in China pursuing the "one-stop-shop" approach in China, despite the fact that the companies employing this technique typically fail.

If I had not been aware of this fact, it is inconceivable that I could have ever formed a high-quality LED Lighting brand that offers excellence in every lighting category.

Think about it.

The industry norm for lighting failures is 3% on average; **YES** failure rates average less than 1%, regardless of the category—impressive to say the least.

This is entirely the result of partnerships with category-specific suppliers, who are in the top of their respective categories, and who met and exceeded every one of our **10 Keys of Success.**

Certainly, most of our supplier-partners offer products within other LED lighting categories. However, our business model calls for the best products made by the best factory in each lighting category.

As we continue to evolve, we maintain our dedication to aligning ourselves with only the very best category-specific LED lighting supplier-partners.

Chapter 12

The Process of Manufacturing LEDs

Consider the fact that over 90% of the world's lighting is **entirely** manufactured or assembled in China, including but definitely not limited to lighting giants such as GE, Philips and Sylvania. Obviously, every single component used in the manufacturing process is available in China. Not to mention, all the equipment, expertise, and labor is located there as well.

Most international LED lighting companies use Chinese factories and labor to manufacture and assemble their products. Therefore, it is important to be aware that the majority of LED lighting products that are labeled "Made in America" are often made of Chinese components and then assembled in the U.S. using Chinese equipment.

Many Chinese companies are fully aware of America's desire to buy "American Made" goods, which has led to the formation of Chinese-owned companies that operate as U.S.-based businesses.

Or, in some cases, the Chinese will partner with U.S-based companies to manufacture and or assemble here in the U.S. (this is not "always" the case, but it is "very often" the case).

I cannot tell you how many times I have cracked open an LED lamp that was labeled "Assembled in America" only to find that every component inside was stamped "Made in China."

Many consumers mistakenly think that the biggest difference in the Made in America and Made in China products is always quality. However, Chinese products are generally better due to China's manufacturing experience and expertise, and prices are significantly lower due to their huge economy of scale.

The point is that you should always be aware of what you are producing. When it comes to LED lighting, there can be a huge difference in the quality of the end product. That's primarily why it is now required to put *LED Lighting Facts* (similar to *Nutritional Facts* on food) and to secure Energy Star and DLC listings for products because of the huge variance in quality.

Let's take the process of encapsulation as an example. Encapsulation is simply the process of shaping the LED diodes so that they can attain maximum brightness, efficiency, and light dispersion. The design of the wafer doesn't lend itself well to shapes other than cubes, so that means shaping has to be done in an extra step. The most popular of these methods is encapsulation.

The way it works is by encapsulating the LED bracket with a material epoxy and putting the mould into an oven. As the epoxy solidifies, the LED becomes shaped.

This may sound like a pretty simple process for such a sophisticated and cutting-edge technology. But there is a long list of tiny little things that can go wrong during the encapsulation process that may leave the LED chips flawed or inoperable (and this is just ONE process of many).

If encapsulation is done poorly, you end up with a product that has an improperly aligned LED bracket. And if you bake the mould too hot or for too long, you will stain the bracket yellow from excessive oxidation. Not only that, but you can get all sorts of crazy anomalies like air bubbles, impurities, and light nebulization. Other outcomes can include watermarks, damages, and cracks in the epoxy.

There are so many things that can go wrong in the manufacture of LEDs. It is absolutely essential to commit to a known and trusted source with a FULL commitment to quality control.

Never trust the person who tells you nothing can go wrong.

Early on, I learned to respect the experts who advised to be completely attentive to the manufacturing processes.

I have wisely heeded their advice as to what systems and procedures to put in place to allow YES LED Lighting to establish and maintain the highest quality LED products possible.

Chapter 13
Factory Visit

First Sight:

I would like to describe what a typical factory visit in China is like. First of all, we always scheduled our meetings in advance in order to meet the chairman and/or the directors of the factories.

The larger factories typically vary from 250,000 to 450,000 square feet in size and employ anywhere between 250 and 1,500+ workers, of whom also include the sales staff and administrative support people.

It is important to explain that the workers in these factories actually live within the confines of the factory. The factories provide dormitories that house hundreds, and often thousands, of people. WKK's factories collectively employ and house over 20,000 people.

Young people are pouring into the cities by the hundreds of thousands seeking jobs and have little or no means to afford housing or food on their own, therefore the factories wisely house and provide for their basic needs.

These facilities are typically located inside huge industrial areas that include shops, restaurants, and just about every type of neighborhood service from restaurants to bicycle repair.

A number of these industrial areas have grown so much that they have literally become their own districts or small cities.

It is important to note that not one single YES LED Lighting supplier-partner has closed its doors.

In fact, all of our partner LED factories have at least tripled or quadrupled in size since we began our relationships in 2007.

Upon entering a factory, depending on whom you are going to see, you are escorted to different types of meeting rooms or offices. The Chinese value ceremony and how this plays out depends on the position of the person you are meeting.

If you are meeting one of the young sales associates who work in the showrooms, you will go to a conference table in the showroom. You will be treated politely, shown around the showroom, and then eventually taken on a brief and partial factory tour, after which you will be seated to discuss products and pricing.

If you are meeting one the factory managers you are escorted to either a nice interior office or conference room and served fresh fruit, hot tea, or coffee.

In my case, we would always meet with the chairman and/or directors. In these meetings, you enter huge offices that are typically 1,000 to 2,500 square feet minimum. The owner or chairman will always have the biggest office in the factory in order to demonstrate his power, prestige, and, of course, wealth.

No expense is spared in these offices. They are generally adorned with lavish desks, credenzas, wall units, sofas, overstuffed chairs and recliners, giant coffee tables and, frequently, a good-sized, modern kitchen. I will admit it often took me a few minutes to acclimate myself to these luxurious surroundings.

It is in these offices where the real decisions were made and the deals struck which were instrumental to the development and growth of YES LED Lighting.

The average person who enters these factories would either not think to—or not be able to—reach the leaders of these companies. Thanks to my relationship with Hamid and WKK LTD., these doors were kindly opened to me.

I can vividly remember my first factory meeting. When we arrived, literally, a red carpet was rolled out, and we walked through a gauntlet of applauding factory workers who bowed as we entered the factory. Next, we were escorted through the facility and into an office that would make Donald Trump feel right at home. Waiting at the door to greet us was a young Chinese man who introduced himself as Marcus, although his Chinese name was Po.

I should mention that Chinese business people use English names for the sake of simplicity as many Chinese names are very difficult to pronounce.

In excellent English, he invited us to take a seat. Once we took our seats, he asked us if we would like to have water, tea, coffee, or a soft drink. We enjoyed several minutes of entertaining small talk before the real conversation began.

As a businessman who had been in countless meetings and

conference room scenarios, I considered myself relatively adept at negotiations. Well…let me tell you something…the Chinese *invented* business negotiations.

At the ripe old age of 38, Po was the chairman of a $350 million-dollar-a-year company that was growing at rate of more than 15% per year.

Over the years, I have become quite close to Po and others like him, and I hold their friendships quite dear. The people whom I am associated with in China have integrity, character, and a sense of honor that extends to their employees and manifests itself in the goods they produce. They are unique and as admired in a country of nearly 1.4 billion people as Bill Gates, Mark Zuckerberg, and Mark Cuban are here in the United States.

My enduring relationships with Po and others of his caliber have been key to our success in building the **YES** brand.

The Factory Floor

One of the most critical places to visit is the factory floor, as viewing the production lines and processes is extremely important. It is crucial to know what you should find as well as what you should not.

It is advisable to notice what type of equipment is being used, how well it is maintained, how old it is, and how the factory is organized to maximize production.

These things have everything to do with the technology

utilized, the quality of the goods that are produced, and the attention given to production efficiency.

Important information can be gathered by simply writing down the names and model numbers of the equipment seen on the production floors. One can learn who made the equipment, how old it is, and what its service life is. This information lets us know whether the supplier is using the most up-to-date technology in their manufacturing processes.

Lacking a background in manufacturing, there were several things that I would not have thought to consider without guidance.

There has always been and always will be technological advances in the equipment used in every area of manufacturing. There are actually machines that manufacture themselves!

Therefore, we are exceedingly careful to ensure **YES** supplier-partners utilize only the most up-to-date technologies in the manufacturing processes.

Better technology offers better products when it comes to designing and building power supplies, encapsulating LED chips, embedding them onto their circuit boards and sealing them into their housing.

Therefore, when we see out-of-date equipment at a factory we automatically eliminate that supplier from consideration as a YES LED Lighting supplier-partner.

The Supply Area

Another important area of any factory is the component supply department. The materials and components used in production are received, stored, and funneled into the factory from this area. In the case of LED lighting, this room can contain either LED diodes already encapsulated, or raw, pre-capped LEDs to be placed on the circuit boards.

You will also find circuit boards, power supplies (a.k.a. drivers) and the components used to assemble them, along with wiring harnesses, aluminum housings, and polycarbonate lens covers which eventually house the finished lamp.

It is important to take the time to tour these areas, as they speak volumes about the organization of the factory. The necessity of a clean and organized supply department is obvious as it is the table from which the factory eats.

Strict organization is paramount, as mistakes, delays, or contamination of the numerous essential components can easily lead to major issues. These can include incorrect components getting into the production lines, contaminated products assembled from contaminated parts, and other problems that result in serious production delays.

It is also important to verify the origin of the components. It is commonly known in the LED industry that most of the high-quality LED components originate in the first tier countries such as the United States, South Korea, Japan, and Taiwan.

It is also known that electronic components made in China are not the same quality as the components manufactured in the above-mentioned countries, and that the components from the first tier countries are often copied.

Therefore, we take the added precaution of verifying the authenticity of the electronic components used in producing our products to ensure against the use of counterfeit parts.

Counterfeiting is a huge business in China and it takes place at virtually every level. It is not uncommon for corruption to take place at even the largest factories in China.

A corrupt supply manager could insert counterfeit products into the supply chain.

The quality of the counterfeiting can be perfect down to the last detail. I have personally witnessed senior level experts from Texas Instruments, Advanced Micro Devices, Sony, Samsung, and other companies inspect counterfeit components, unable to tell the difference with the naked eye.

The prevention process is relatively simple, as we have our own quality control staff that verifies the precise source of all our product components. We also run random diagnostic checks to ensure that our meticulous specifications are met.

This procedure plays a major role in ensuring that our products remain among the highest quality in the lighting industry.

Quality Control

For obvious reasons, it's important to be diligent in assessing a factory's capability, capacity and concern for quality control.

Although we here at YES have our own guidelines and quality control staff to verify the quality of our products, it is important that the factory also has its own inherent interest in quality, as well as the systems in-place to ensure it.

I quickly learned that the level of in-house quality control is directly linked to the factory's ability to produce high-quality products.

A factory with inadequate quality control typically:

- Arrogantly assumes that its production capability is so good that there will be no mistakes.

- Attempts to cut cost by not paying for this vital function.

- Attempts to get its products shipped more quickly by avoiding this key element of the manufacturing process.

Regardless of the reason, the best way to put your company in peril of failure is to ignore the necessity for strict and consistent quality control.

Every **YES** product undergoes a five-to-seven-day quality assurance test, which includes the lamps or fixtures being turned on and off at least 200 times at varying intervals.

An easy way to determine if a factory has the right quality controls is to ask them what their product production volumes are. Once we know this, we ask them to show us their testing area along with their reports from previous orders.

If the testing area is NOT large enough to support testing their stated production and the reports aren't available then the supplier is either being deceptive about either their production capacity or their quality controls. Neither mistruth is acceptable.

Reviewing a factory's testing protocols will reveal volumes about their dedication to quality control or about the potential shortcomings of their existing in-house quality control measures. We accept no shortcuts in the testing phase. We have established our own stringent testing protocols that call for the generation of reports that provide specific dates of testing, along with performance tolerances and failure rates.

Chapter 14

Purchase Contracts

Once the Lighting Fair had concluded, my factory tours were finished, and my understanding of business deals in China was on solid footing, I was finally ready to place my first product orders (known in China as P.I.s, or Purchase Invoices).

As I mentioned earlier, it is essential to note every single detail in the purchase contract. The Purchase Invoice is considered the contract and lists all the specific terms of your order. No detail is too small or insignificant.

It is important to note that IF any detail is omitted, it does not exist in the case of a problem.

The exact specification of each product must be listed in this contract, along with the precise type of components that are to be used. The warranty terms are obviously critical and must be noted on the P.I., otherwise no warranty will be in effect.

Even the cost of loading the container and the transportation cost to get the containers to the port must be listed. The P.I. must also specify who will be responsible for the cost of getting the containers through customs and aboard the vessel that will transport them to their foreign destination.

There are many significant details which can and will affect bottom line cost and subsequent profits, so it is critical to be familiar with the numerous and complex factors involved in moving goods from China to the U.S.

Over time, and through the friendships and partnerships formed during my many visits to China, I have learned many ways to minimize costs and protect our interests.

It is important to mention, that to this day every YES LED Lighting contract with our supplier-partners in China has been honored.

Chapter 15

What Sets YES LED Lighting Apart?

I quickly noticed in my initial research that none of the major U.S. lighting suppliers produced what could be considered a full line of interior and exterior LED lighting. I found this fact to be very interesting and it took me a while to figure out why this was the case.

The reason is that traditional light bulbs are a commodity just like food, clothing, and tires, and is therefore integral to satisfying people's wants and needs.

Every month, billions of light bulbs are discarded and replaced around the world. Therefore, it is no wonder that Philips, GE, Sylvania, and similar companies remain immersed in the interior lighting segment of the industry, a huge commodity business.

For these companies, retooling to enter a different market, like exterior lighting, is a slow, expensive, and painstaking process. Much like adjusting the course of a huge ship, it takes a long time to adjust the rudder, alter the direction, and regain cruising speed.

The major lighting companies have utilized supplier-partners that have been traditional bulb manufacturers for decades. These factories do not have neither the necessary tools nor equipment to produce exterior LED lighting.

I made similar observations while visiting dozens of other factories in China. These visits helped me to identify major gaps in the LED lighting industry.

After touring a number of top suppliers, taking care to acknowledge their core strengths and products, I noticed an obvious pattern: each factory was especially proficient in one particular category of LED lighting, despite also producing other products.

Using an earlier example, our LED tube manufacturer also made LED street lighting. However, it was obvious and easy to notice where their efforts were focused. When walking the factory production floors I could see that 600 workers were assembling LED tubes in a 250,000 square foot area while 100 workers were assembling LED streetlights in a 25,000 square foot area. Likewise, our streetlight supplier had 500+ workers assembling LED streetlights while about 60 workers were assembling tubes.

This critical observation remains a major key to our continued success and growth.

Why? The answer is simple.

I chose to partner YES LED Lighting with the best supplier in each individual category of LED lighting. And the process of choosing supplier-partners and testing the various products took several years to complete to my satisfaction.

Many U.S.-based companies have created major problems for themselves by failing to understand the importance of choosing partners who are leaders in their LED category.

As our Chinese relationships have grown and strengthened over time, many of our Chinese partners have told me (off-the-record, of course) that it was wise to take the time to identify and partner with the best of the best in every category.

To date, we continue to expand and improve our full range of superior-quality LED lighting.

Chapter 16

The Take-Away:
What I Have Learned

China manufactures the vast majority, over 92%, of the world's lighting products. This, of course, includes LED lighting. It's truly incredible how vast and multi-faceted manufacturing in China really is.

As I toured factories and was allowed to actually, by request, participate in the manufacturing and assembly process, I noticed that different factories engaged in many different levels of manufacturing.

Some factories had the ability to manufacture lighting from absolute scratch. In these cases, they possessed the equipment to pour the aluminum and or plastics into molds to produce the housings. They also had the equipment to encapsulate LEDs, embed them on the circuit boards that are then attached to the power supplies and secure and seal them into the aluminum housing to complete the process of manufacturing a lamp.

Others I toured were assembling lamps from prefabricated components. I learned that this is due to an economy of scale and, if properly monitored, made little tangible difference.

In some cases factories actually purchased some of these processes or components already prepared and/or pre-formed. My point here is that there are many variations as to how factory production takes place in China.

I also learned early on that China's major asset in product production is their ability to provide everything at a lower cost, due to their gigantic economy of scale. Without a doubt, China has systematically established itself as the world's unparalleled manufacturing leader.

Another important thing I learned is that there are two types of products made in China. The first type is known as "Chinese Domestic Products." These products are made using Chinese-made components. The quality of these products is very low in comparison to the second type of products, known as "International Market Products." What sets the two types apart is, of course, the quality of components that are used.

YES products are manufactured utilizing only international-quality product components. The top quality components (as mentioned earlier) are produced by companies such as Texas Instruments, Samsung, Sony, Philips, and other top tier electronics providers, originating from factories located in the United States, South Korea, Japan, and Taiwan.

Having a keen insight into how the LED industry works is one of the main contributors to the success of the YES brand.

None of this would be possible without the support and ongoing mentorship of everyone from my friend Herbert who introduced me to Hamid, to Peter and his invaluable industry knowledge and translation skills. Without these people, I'm confident that the **YES** brand would not be what it is today and to them and many others I want to express my deepest gratitude.

One of the most important things I've learned is the value of forging high quality business relationships.

It was these key people who illuminated the path for YES LED Lighting to achieve top ratings in every lighting category.

A crucial element to our success was that I was able to tour the actual factories and see first-hand what worked and what didn't (and why).

I had the great fortune of being able to handpick the elements of success along the way.

This fact, along with our *strict* adherence to the 10 Keys to Success, will absolutely ensure the future growth and success of YES LED Lighting.

It is for this reason that I am eternally grateful to Hamid for practically handing the 10 Keys to me on a silver platter.

By creating and nourishing powerful alliances, we have turned great ideas into amazing things. As someone who was given a great deal of help on my journey, I see myself in position to share with others a ton of valuable information that was NOT easy to come by.

One of the reasons I'm so passionate about LED lighting is its potential to change the world. Its numerous advantages, along with its eco-friendly design and build, make it the perfect environmentally responsible solution to lighting our world.

As the Founder and President of YES LED Lighting, it is my mission to be a steward for the planet by always pushing for better efficiency, while at the same time drastically reducing the

Earth's carbon footprint.

I consider it an honor to educate people on the vast benefits of converting to LED lighting technology.

The benefits of switching the entire world over to LED lighting are incomprehensible. According to the U.S. Department of Energy, "Widespread use of LED lighting has the greatest potential impact on energy savings in the United States [alone]. By 2027...[LED's could allow for] a total savings of more than $300 billion at today's energy cost."

Chapter 17

Jody's Vision of LED Lighting's Future

Take a look around at the mass production of light bulbs and ask yourself, "What would it be like if light bulbs and light fixtures lasted ten to fifteen years?"

The sad truth is that very few seem to consider the fact that every time someone throws away a light bulb or fixture, it ends up sitting in a landfill for a very long time.

While on the other hand LEDs, along with boasting other amazing attributes, are fully recyclable.

I am hopeful that soon people everywhere will become more conscious of the environment and our responsibility to protect it.

The incentives to switch to LED lighting are already there: it really doesn't require convincing people—only informing them.

I truly believe that as time passes we will see continue to see more and more LED "devotees."

As new advancements continue to drive down the cost, and the performance capabilities of this amazing technology improve, the world will inevitably see LED lighting take its rightful place at the forefront of lighting technology.

I am certain that there will be increased governmental and

regulatory resistance to the continued use of antiquated and wasteful lighting technology.

In the coming years there will be mounting pressure for consumers to convert to superior and more environmentally responsible LED lighting technology.

I think the biggest push will come from social pressure. We'll soon be at the point where those who are still using antique lighting technology may be considered either lazy or crazy.

Furthermore, because LED lighting can be linked to computer technology, there is no end to the amazing things you can do with them. We are on the verge of having some incredibly futuristic lighting technology.

In the business place, LED lights could be used to transfer data as you move through the office. The lights will then automatically have your active data ready at any workstation you use.

In the not-too-distant future, LED lighting will be analogous to cell phones in the sense that the main function of the device ends up being only a small portion of what the device is actually used for.

One of the primary focuses of LED technology is connectivity. By giving the electronic chips networking capability, we will start seeing data transfer in ways most people have never dreamed of.

For example, what if the headlights of the car in front of you could see upcoming hazards in the road and could automatically transmit that data to your car, giving you a great deal of warning time. LEDs have already been demonstrated to be able to do all sorts of amazing things like this. It's all in the realm of possibility.

What really excites me is their potential in everyday lighting use. Currently evolving LED technology has the ability to communicate to virtually any other device. Imagine driving home from the office and your house lights turn on right as you pull into the driveway. Or you could set it so you never again have to turn on a light switch as the room you enter automatically lights up.

You could literally have all of the lights in your home or office set to various mood presets that change automatically based on the setting. Imagine creating a preset for parties that displays colorful, vibrant lights that change and pulse to the beat of the music. When the party's over and it's time to clean up, simply switch the lighting scheme to a crisp, clean, and bright white light. And when it's time to wind down for the evening? Set a nice, warm glow to help you relax.

As well, the programmability of LEDs is virtually infinite. What if, for example, you went on vacation for two weeks and you didn't want to attract the attention of potential thieves and vandals? You could basically record your average lighting habits and then replay them for the duration of your vacation, effectively giving the illusion that the home is occupied and being used normally.

I've only just offered a few examples to help *illuminate* what's possible through the application of LED lighting. I'm continually impressed at the technology and how far it's come in such a short amount of time. I look forward to the coming years when the cost of a standard LED lamp or fixture will actually be the same cost or even less expensive than traditional lighting.

I hope that the experiences and information that I have shared have been helpful in furthering your understanding of LED technology and its benefits.

I also hope you gained some insight on what it takes to build and maintain a successful LED lighting company.

YES LED Lighting continues to grow and establish itself as a company with remarkable products and extremely rare industry knowledge.

Our goal will always be to ensure that our customers have the utmost confidence in our ability to meet their lighting needs. We realize that our customers will always be the most important judge of the quality of our products, and we will always strive to deliver the best of the best LED lighting to them.

We are truly on the cusp of a lighting revolution, and YES LED Lighting is determined to remain at its forefront.

If you are interested in learning more, you can visit our website, www.YESLEDLighting.com.

Thank you for taking the time to read my book.

Jody Cloud

Appendix 1

Action Pictures

Another day at the office

Taking inventory

Overseeing delivery

Appendix 2

Samples of Our Work

Commercial / office space

Recreational facility

Hospitality

Car dealership

Appendix 3

Glossary of Terms

Delivered Light:

The amount of light a lighting fixture or lighting installation delivers to a target area or task surface, measured in foot candles (fc) or lux (lx).

Diffuser:

An object with irregularities on a surface causing scattered reflections for the purpose of increasing light flow.

Directional Light Source:

A light source that emits light only in the direction it is pointed or oriented.

Driver:

Electronics used to power illumination devices.

Efficacy:

The light output of a light source divided by the total electrical input to that source, expressed in lumens per watt (LM/W).

ELV-Type Dimmer:

An electronic low voltage dimmer, used to dim LED lighting fixtures with electronic transformers.

Flux / Luminous Flux:

Luminous flux is the measure of the perceived power of light, adjusted to reflect the varying sensitivity of the human eye to different wavelengths of light.

Ghosting:

An effect that occurs when lighting fixtures in the 'Off' state faintly glow as a result of residual voltage in the circuit.

Heat Sink:

A part of the geo-thermal system that conducts or conveys heat away from sensitive components, such as LEDs and electronics.

High Power LED:

A high power LED, sometimes referred to as a power LED, is one that is driven at a current of 350 mA or higher.

Illuminance:

The intensity of light falling on a surface area, if the area is measured in square feet, the unit of illuminance is measured in footcandles (fc). If the area measured in square meters the illuminance is measured in lux (lx).

Kelvin Temperature:

Term and symbol (K) used to indicate the comparative color appearance of a light source when compared to a theoretical blackbody. Yellowish incandescent lamps are 3000 K. Fluorescent light sources range from 3000 K to 7500 K and higher.

LED Array:

An assembly of LED packages on a printed circuit board or substrate, possibly with optical elements and additional thermal, mechanical, and electrical interfaces that are intended to connect to the load side of an LED driver.

LED Chip:

The light producing semiconductor device that may or may not be incorporated into an LED.

LED Driver:

An electronic circuit that converts input power into a current source—a source in which current remains constant despite fluctuations in voltage. An LED driver protects LEDs from normal voltage fluctuations, over-voltages, and voltage spikes.

LED Light Engine:

An integrated assembly comprised of LEDs or LED arrays, LED driver, and other optical, thermal, mechanical, and electrical components.

LED Luminaire:

A complete lighting unit consisting of LED-based light emitting elements and a matched driver together with parts to distribute light, to position and protect the light elements, and to connect the unit to a branch circuit. The LED light emitting elements may take the form of LED packages (components), LED arrays (modules), LED light engines, or LED lamps. The LED luminaire is intended to connect directly to a branch circuit.

Light Emitting Diode:

A Light Emitting Diode (LED) is a solid-state semiconductor device that converts electrical energy directly into light. On its most basic level, the semiconductor is comprised of two regions. The p-region contains positive electrical charges while the n-region contains negative electrical charges. When the voltage is applied and current begins to flow, the electrons move across the n region into the p region. The process of an electron moving through the p-n region releases energy. The dispersion of this energy produces photons with visible wavelengths.

Power Factor:

The active power divided by the apparent power. Defines what percentage of power is actually light and that which is heat.

Radiant Flux:

The total energy emitted by a light source across all wavelengths, measured in watts.

SMD:

Surface mounted LEDs.

Solid-State Lighting:

A description of the devices that do not contain moving parts OR parts that can break, rupture, shatter, leak, or contaminate the environment.

UV:

Electromagnetic radiation with wavelength shorter than that of visible light.

Volt:

The term that is used to describe the potential electrical differences between oppositely charged conductors.

Watt:

The unit of electrical power that is used by an electrical device during its operation.

Appendix 4

Frequently Asked Questions

How bright are LED light bulbs?

LED bulbs available for standard fixtures vary in brightness from less than 50 lumens up to about 1200 lumens. The brightest LED bulbs for standard fixtures are the floodlights and spotlights. The brightest of these uses about 25 watts and produces light comparable to a 120-watt incandescent.

The brightest LED bulbs with approximately the same size and shape as ordinary incandescent bulbs produce up to 600 lumens. With a few exceptions, these bulbs are somewhat directional so they are most effective when pointed at the area to be illuminated.

How efficient are LED bulbs compared to incandescent / halogen bulbs?

The efficacy of the newer LED light bulbs is more than five times higher than comparable incandescent bulbs. In other words, LED light bulbs use only about 20% as much electricity to produce the same amount of light. However, because LED bulbs direct a larger percentage of light where it is needed, in many applications they are as much as ten times as effective as incandescent bulbs, reducing energy use by 90%.

Are bulbs with more LEDs brighter than bulbs with less?

The number of LEDs is not the determining factor of bulb brightness. Different types of LEDs vary greatly in size and light output. The most accurate indicators of the brightness of LED bulbs are the measured lumens or lux. Lumens measure the total amount of light output from a bulb. Lux measures how bright the light is on a surface at a specified distance.

The brightest LED bulbs with approximately the same size and shape as ordinary incandescent bulbs produce up to 600 lumens. With a few exceptions these bulbs are somewhat directional so they are most effective when pointed at the area to be illuminated.

How does the brightness of LED lighting compare to incandescent lighting?

LED light bulbs are much brighter than incandescent or halogen bulbs of the same wattage, but LED bulbs are not available in very high wattages. Thus, when replacing incandescent or halogen lamps with LED lamps, more LED lamps are often needed. For example, to replace one 100-watt incandescent bulb you may need two 5-watt or 6-watt LED bulbs. Although you have more bulbs you are still using 85% less electricity.

What do cool white and warm white mean, and what is CCT?

The Color Correlated Temperature (CCT) is given in the description of each of our white LED bulbs. The color (CCT) of our white bulbs ranges from a warm yellow white (2700 K) to a cool blue white (7000 K).

By comparison, a typical incandescent bulb has a CCT of 2800 K. A typical halogen is a bit higher, maybe 3500 K. Daylight white is 4500K and a cool white fluorescent is 6000 K or more.

The human eye adapts to background light so that even a daylight white bulb will look slightly blue in a room illuminated mainly with incandescent bulbs. Similarly, an incandescent bulb will look very yellow or even orange in midday sunlight.

What is the difference between a floodlight and a spotlight?

LED spotlights output a narrower beam of light, typically less than 45 degrees wide. Most of the light from a spotlight is concentrated onto a relatively small area producing a bright spot. LED floodlights output a wider beam of light, up to 120 degrees, so the light from a floodlight is spread out over a much larger area.

Because the light is more concentrated, a spotlight will appear brighter than a floodlight but only within its narrower beam. A spotlight is more suited to illuminating objects and a floodlight is more suited to illuminating areas.

Appendix 5

Facts and Key Players

One of my goals with this book is to leave the reader feeling well informed. This section talks more about the general landscape of manufacturing, including all of the logistics and safety measures that go along with it.

The companies and institutions in this section are especially integral to the LED lighting industry and serve to paint a more complete picture of what's involved in ensuring the high caliber products offered by YES LED.

Every single product manufactured by YES LED Lighting must adhere to all of these standards and more in order to truly soar above the competition. So far we've been able to achieve top ratings in every single lighting category.

None of this would be possible without a strict compliance with the following rules, regulations, and protocols. By now you're probably aware of the degree of care that the YES brand has when it comes to everything related to quality and safety.

The Occupational Safety and Health Administration (OSHA) is a U.S. agency dedicated to assuring safe and healthful working conditions for working men and women by setting and enforcing standards and by providing training, outreach, education, and assistance.

OSHA has always been the key agency protecting the rights of employees and setting clear responsibilities for the employers that hire them.

The OSH Act gives workers the right to safe and healthful working conditions. It is the duty of employers to provide workplaces that are free of known dangers that could harm their employees. This law also gives workers important rights to participate in activities to ensure their protection from job hazards. The following is a list of these rights:

Workers' Rights under the OSH Act

- File a confidential complaint with OSHA to have their workplace inspected.

- Receive information and training about hazards, methods to prevent harm, and the OSHA standards that apply to their workplace. The training must be done in a language and vocabulary workers can understand.

- Review records of work-related injuries and illnesses that occur in their workplace. Receive copies of the results from tests and monitoring done to find and measure hazards in the workplace.

- Get copies of their workplace medical records.

- Participate in an OSHA inspection and speak in private with the inspector.

- File a complaint with OSHA if they have been retaliated against by their employer as the result of requesting an inspection or using any of their other rights under the OSH Act.

- File a complaint if punished or retaliated against for acting as a "whistleblower" under the additional 21 federal statutes for which OSHA has jurisdiction.

A job must be safe or it cannot be called a good job. OSHA strives to make sure that every worker in the nation goes home unharmed at the end of the workday, the most important right of all.

In addition to worker's rights, employees have responsibilities to ensure that their workplace is sufficiently safe for performing any type of work.

Employer Responsibilities under the OSH Act

Employers have the responsibility to provide a safe workplace. Employers MUST provide their employees with a workplace that does not have serious hazards and must follow all OSHA safety and health standards.

Employers must find and correct safety and health problems. OSHA further requires that employers must try to eliminate or reduce hazards first by making feasible changes in working conditions – switching to safer chemicals, enclosing processes to trap harmful fumes, or using ventilation systems to clean the air are examples of effective ways to get rid of or minimize risks – rather than just relying on personal protective equipment such as masks, gloves, or earplugs.

Employers MUST also:

- Prominently display the official OSHA poster that describes rights and responsibilities under the OSH Act. This poster is free and can be downloaded from www.osha.gov.

- Inform workers about hazards through training, labels, alarms, color-coded systems, chemical information sheets, and other methods.

- Train workers in a language and vocabulary they can understand.

- Keep accurate records of work-related injuries and illnesses.

- Perform tests in the workplace, such as air sampling, required by some OSHA standards.

- Provide hearing exams or other medical tests required by OSHA standards.

- Post OSHA citations and injury and illness data where workers can see them.

- As of January 1, 2015, notify OSHA within 8 hours of a workplace fatality or within 24 hours of any work-related inpatient hospitalization, amputation, or loss of an eye.

- Not retaliate against workers for using their rights under the law, including their right to report a work-related injury or illness.

UL is a global independent safety science company with more than a century of expertise innovating safety solutions from the public adoption of electricity to new breakthroughs in sustainability, renewable energy, and nanotechnology.

The company is wholly dedicated to promoting safe living and working environments. UL helps safeguard people, products, and places in important ways, facilitating trade and providing peace of mind.

UL certifies, validates, tests, inspects, audits, and advises and educates. They provide the knowledge and expertise to help navigate growing complexities across the supply chain from compliance and regulatory issues to trade challenges and market access.

Intertek

The ETL Listed Mark is proof that your product has been independently tested and meets the applicable published standard.

Intertek's ETL Mark was born into a culture of innovation. It was in Thomas Edison's lighting laboratories where it all began, and to this day we still breathe the same air of innovation, safety, and quality. We also understand a manufacturer's need to get new products to market quickly to achieve the greatest success, therefore they have built speed, responsiveness and urgency, into their processes. Their commitment to helping customers gain the certifications they need quickly and efficiently has never been greater.

The ETL Mark is proof of product compliance to North American safety standards. Authorities Having Jurisdiction (AHJs) and code officials across the U.S. and Canada accept the ETL Listed Mark as proof of product compliance to published industry standards. Retail buyers accept it on products they're sourcing. And every day, more and more consumers recognize it on products they purchase as a symbol of safety.

Today, the ETL Mark is the fastest growing safety certification in North America and is featured on millions of products sold by major retailers and distributors every day.

The TÜV Rheinland Group is a leading provider of technical services worldwide.

Founded in 1872 and headquartered in Cologne, the Group employs 19.320 people in 500 locations in 69 countries. It generates annual revenues of € 1.7 billion.

The Group's mission and guiding principle is to achieve sustained development of safety and quality in order to meet the challenges arising from the interaction between man, technology and the environment.

As an international service group, they [TÜV] document the safety and quality of new and existing products, systems and services.

A Program of the U.S. DOE

The U.S. Department of Energy (DOE) created the LED Lighting Facts program to assure decision makers that the performance of solid-state lighting (SSL) products is represented accurately as products reach the market. Sensitive to the setbacks that plagued consumer adoption of other new technologies, DOE developed the LED Lighting Facts program to manage user expectations and prevent the exaggerated performance claims that are often prevalent with new technologies.

Becoming an LED Lighting Facts partner requires a commitment to supporting improvement of the quality of SSL products, as well as using the LED Lighting Facts labels and logos according to program guidelines. Each partner must pledge to honor this commitment and uphold program goals specific to each partner type.

Partners with this program will be required to print a physical label on LED lighting products that gives the consumer specific information about the efficiency, light output, and even color of the LED lighting product (ranges from 'cool' to 'warm'). Although there are currently four versions of the label, each with optional metrics, the next page has a good example of what one looks like.

lightıng facts^{CM}

A Program of the U.S. DOE

Light Output (Lumens)	**336**
Watts	**6**
Lumens per Watt (Efficacy)	**60**

Color Accuracy Color Rendering Index (CRI)	**86**

Light Color
Correlated Color Temperature (CCT)

3013 (Bright White)

Warm White	Bright White	Daylight
2700K 3000K	4500K	6500K

All results are according to IESNA LM-79-2008: *Approved Method for the Electrical and Photometric Testing of Solid-State Lighting.* The U.S. Department of Energy (DOE) verifies product test data and results. Products qualified under the DOE ENERGY STAR® program have the ENERGY STAR mark on this label.

Visit www.lightingfacts.com for the *Label Reference Guide*.

Registration Number: DQGW-1C365E

Model Number: 6 Watt Dimmable LED Bulb (A19)

Type: Replacement lamps

Anatomy of the Label

Light Output/Lumens

- Measures light output. The higher the number, the more light is emitted.

- Reported as "Total Integrated Flux (Lumens)" on LM-79 test report.

Watts

- Measures energy required to light the product. The lower the wattage, the less energy used.

- Reported as "Input Power (Watts)" on LM-79 test report.

Lumens per Watt/Efficacy

- Measures efficiency. The higher the number, the more efficient the product.

- Reported as "Efficacy" on LM-79 test report.

IESNA LM-79-2008

- Industry standardized test procedure that measures performance qualities of LED luminaires and integral lamps.

- Allows for a true comparison of luminaires regardless of the light source.

LED Lumen Maintenance

- Listed as a percentage, this metric estimates the amount of light the LED light source is projected to emit at 25,000 hours at a given ambient test temperature, compared to its initial light output. This percentage is based on LM-80, in-situ performance, and TM-21 projections.

Model Number

- Unique manufacturer's model number for the product.

Type

- Specific type of solid-state lighting fixture.

Brand

- The brand under which each product is available.

Color Rendering Index (CRI)

- Measures color accuracy.

- Color rendition is the effect of the lamp's light spectrum on the color appearance of objects.

Correlated Color Temperature (CCT)

- Measures light color.

- "Cool" colors have higher Kelvin temperatures (3600–5500 K).

- "Warm" colors have lower color temperatures (2700–3500 K).

Product Warranty

- If a label indicates that the product has a warranty, a URL with more information has been verified by the program and is available from the product summary.

Appendix 6

Industry Certifications

It is very important to understand the function and meaning of the manufacturing certifications of products made in China. The basis of the certifications is to ensure that the products are safe. To reach the high standard of safety required in countries such as the U.S., it is necessary for designers and manufacturers to create and produce products using high-quality components to assure safety standards.

Therefore, if a product does not have the necessary certifications, such as UL (Underwriters Laboratories) and DLC (Design Lights Consortium), you can be sure that the quality of the product is not up to par.

OSHA (Occupational Safety and Health Administration) has approved laboratories such as UL and ETL (Environmental Technology Laboratory) to test products and issue safety certifications.

The purpose of these certifications is to ensure that the products inspected by the designated laboratories meet certain specific and stringent safety standards. The inspection itself is a process to ensure that the product as a whole and each individual component meets or exceeds the standards set forth by OSHA.

In the U.S., depending on the item, insurance companies typically stipulate OSHA standards be met and maintained.

For example, if a building is insured against fire loss, an insurance company would demand that all electrical components, including wiring and switches, be manufactured and certified to meet OSHA standards.

It is absolutely necessary to verify the authenticity of Chinese product certifications, as we have received countless fake UL, ETL, EnergyStar, and DLC documents.

Also, many of these disreputable companies will state that they have certifications that they do not actually have as they anticipate obtaining the certifications after receiving an order…or not.

UL (Underwriters Laboratories) is a safety consulting and certification company headquartered in Northbrook, Illinois. It maintains offices in 46 countries. **UL** was established in 1894 and has participated in the safety analysis of many of the last century's new technologies, most notably the public adoption of electricity and the drafting of safety standards for electrical devices and components.

UL provides safety-related certification, validation, testing, inspection, auditing, advising, and training services to a wide range of clients, including manufacturers, retailers, policymakers, regulators, service companies, and consumers.

UL is one of several companies approved to perform safety testing by the U.S. federal agency Occupational Safety and Health Administration (OSHA). OSHA maintains a list of approved testing laboratories, which are known as Nationally Recognized Testing Laboratories.

Intertek is an OSHA (Occupational Safety & Health Administration) recognized NRTL (Nationally Recognized Testing Laboratory) and is accredited as a Testing Organization and Certification Body by the Standards Council of Canada.

Intertek is one of the world's largest testing, inspection and certification companies. They have more than 60 laboratories across North and South America, Europe and Asia to deliver electrical safety testing and certification for products.

The **DesignLights Consortium**: (**DLC**) is a project of Northeast Energy Efficiency Partnerships (NEEP), a regional non-profit founded in 1996 whose mission is to serve the Northeast and Mid-Atlantic to accelerate energy efficiency in the building sector through public policy, program strategies and education. The **DLC** promotes quality, performance, and energy efficient commercial sector lighting solutions through collaboration among its federal, regional, state, utility, and energy efficiency program members; luminaire manufacturers; lighting designers, and other industry stakeholders throughout the U.S. and Canada.

Energy Star (trademarked *ENERGY STAR*) is an international standard for energy efficient consumer products originated in the United States. It was created in 1992 by the Environmental Protection Agency and the Department of Energy. Since then,

Australia, Canada, Japan, New Zealand, Taiwan, and the European Union have adopted the program. Devices carrying the Energy Star service mark, such as computer products and peripherals, kitchen appliances, buildings, and other products, generally use 20–30% less energy than required by federal standards. In the United States, the Energy Star label is also shown on the EnergyGuide appliance label of qualifying products.

The **IP Code**, **International Protection Marking**, IEC standard 60529, sometimes interpreted as **Ingress Protection Marking**,[1] classifies and rates the degree of protection provided against the intrusion (including body parts such as hands and fingers), dust, accidental contact, and water by mechanical casings and electrical enclosures. It is published by the International Electrotechnical Commission (IEC).

The standard aims to provide users with more detailed information than vague marketing terms such as *waterproof*. The digits (characteristic numerals) indicate conformity with the conditions summarized in the tables below. Where there is no protection rating with regard to one of the criteria, the digit is replaced with the letter *X*.

First Digit: Solids

The first digit indicates the level of protection that the enclosure provides against access to hazardous parts (e.g., electrical conductors, moving parts) and the ingress of solid foreign objects.

Level	Object size protected against	Effective against
0	Not protected	No protection against contact and ingress of objects
1	>50mm	Any large surface of the body, such as the back of the hand, but no protection against deliberate contact with a body part.
2	>12.5mm	Fingers or similar objects.
3	>2.5mm	Tools, thick wires, etc.
4	>1mm	Most wires, screws, etc.
5	Dust Protected	Ingress of dust is not entirely prevented, but it must not enter in sufficient quantity to interfere with the satisfactory operation of the equipment; complete protection against contact.
6	Dust Tight	No ingress of dust; complete protection against contact.

Second Digit: Liquids

Protection of the equipment inside the enclosure against harmful ingress of water.

Level	Object size protected against	Effective against
0	Not protected	-
1	Dripping water	Dripping water (vertically falling drops) shall have no harmful effect.
2	Dripping water when tilted up to 15°	Vertically dripping water shall have no harmful effect when the enclosure is tilted at an angle up to 15° from its normal position.
3	Spraying water	Water falling as a spray at any angle up to 60° from the vertical shall have no harmful effect.
4	Splashing water	Water splashing against the enclosure from any direction shall have no harmful effect.
5	Water jets	Water projected by a nozzle (6.3mm) against enclosure from any direction shall have no harmful effects.
6	Powerful water jets	Water projected in powerful jets (12.5mm nozzle) against the enclosure from any direction shall have no harmful effects.
7	Immersion up to 1m	Ingress of water in harmful quantity shall not be possible when the enclosure is immersed in water under defined conditions of pressure and time (up to 1 m of submersion).
8	Immersion beyond 1m	The equipment is suitable for continuous immersion in water under conditions which shall be specified by the manufacturer. Normally this will mean that the equipment is hermetically sealed. However, with certain types of equipment, it can mean that water can enter but only in such a manner that it produces no harmful effects.

For an expanded IP Rating Reference Chart, please visit http://www.dsmt.com/resources/ip-rating-chart

For example, an electrical socket rated IP22 is protected against insertion of fingers and will not be damaged or become unsafe during a specified test in which it is exposed to vertically or nearly vertically dripping water. Another example is the Sony Xperia Z Ultra, one of the first cellular phones to be IP-rated; it is rated at IP58 and marketed as "dust resistant" and can be "immersed in 1.5 meters of freshwater for up to 30 minutes". Other examples include the Samsung Galaxy S5 and the Fluke27 II & 28 II Digital Multimeters, rated at IP67, also with a high degree of resistance to dust and water. IP22 or 2X are typical minimum requirements for the design of electrical accessories for indoor use.

The definition and aim of the **RoHS** directive is quite simple. The **RoHS** directive aims to restrict certain dangerous substances commonly used in electronics and electronic equipment. Any **RoHS** compliant component is tested for the presence of Lead (Pb), Cadmium (Cd), Mercury (Hg), Hexavalent chromium (Hex-Cr), Polybrominated biphenyls (PBB), and Polybrominated diphenyl ethers (PBDE). For Cadmium and Hexavalent chromium, there must be less than

0.01% of the substance by weight at raw homogeneous materials level. For Lead, PBB, and PBDE, there must be no more than 0.1% of the material, when calculated by weight at raw homogeneous materials. Any **RoHS** compliant component must have 100 ppm or less of mercury and the mercury must not have been intentionally added to the component. In the EU, some military and medical equipment are exempt from **RoHS** compliance.

The **CE mark**, or formerly **EC mark**, is a mandatory conformity marking for certain products sold within the European Economic Area (EEA) since 1985. The **CE** marking is also found on products sold outside the EEA that are manufactured in, or designed to be sold in the EEA. This makes the CE marking recognizable worldwide even to people who are not familiar with the European Economic Area. It is in that sense similar to the FCC Declaration of Conformity used on certain electronic devices sold in the United States.

It consists of the **CE** logo and, if applicable, the four digit identification number of the notified body involved in the conformity assessment procedure.

The **CE** marking is the manufacturer's declaration that the product meets the requirements of the applicable EC directives.

The actual words signified by "**CE**" have been disputed. It is often taken to be an abbreviation of *Conformité Européenne*, meaning "European Conformity". However, "**CE**" originally stood for "*Communauté Européenne*", French for "European Community". In former German legislation, the **CE** marking was called "*EG-Zeichen*" meaning "European Community mark". The **CE** marking is a symbol of free marketability in the European Economic Area (Internal Market).

Bibliography

Anatomy of the Label. (2015). Retrieved September 2, 2015.

Compare: LED Lights vs CFL vs Incandescent Lighting Chart. (n.d.). Retrieved September 1, 2015.

Definition and Description of UL. (n.d.). Retrieved September 1, 2015.

Day, J. (2014). Understanding Watts vs. Lumens for Home Lighting | Today's Homeowner. Retrieved September 1, 2015.

Employee and Worker Rights on pages 2-3 (also in Appendix 6). (n.d.). Retrieved September 1, 2015.

ETL Listed Mark. (n.d.). Retrieved September 1, 2015.

Harris, T., & Fenlon, W. (2015). How Light Emitting Diodes Work. Retrieved September 1, 2015.

Heber, G. (2013, November 27). Environmental Benefits of Using LED Lights. Retrieved September 3, 2015.

How Does a Lightbulb Work? (1992, June 17). Retrieved September 1, 2015.

IP Code. (n.d.). Retrieved September 26, 2015.

K., M. (2013, March 29). Anatomy of an LED Bulb. Retrieved September 3, 2015.

Keeping, S. (2011, September 9). Improving the Efficiency of LED Light Emission. Retrieved September 3, 2015.

Kelly-Detwiler, P. (2015, May 13). Philips And The Future Of LED Lighting. Retrieved September 3, 2015.

LED Encapsulation Process - LEDinside. (2009, July 24). Retrieved September 3, 2015.

Light Bulbs: Incandescent vs. CFL vs. LED. (n.d.). Retrieved September 1, 2015.

Monsanto MV1 - The First Successful Red LED. (n.d.). Retrieved September 1, 2015.

OSHA'S Mission (see Appendix 6 in YES LED Lighting book). (n.d.). Retrieved September 1, 2015.

Schuellerman, D. (2012). LED Inventor Nick Holonyak Reflects on Discovery 50 Years Later. Retrieved September 1, 2015.

Seesmart. (2010). Retrieved September 1, 2015.

T., D. (2012, December 12). Top 10 Benefits of Using LED Lighting. Retrieved September 1, 2015.

TUV Info. (2015). Retrieved September 1, 2015.

The History of the Light Bulb. (2013, November 22). Retrieved September 1, 2015.

Woodford, C. (2008). Diodes and LEDs. Retrieved September 1, 2015.

 LED LIGHTING

Jody Cloud
(305) 586-4800
info@yesledlighting.com
www.YESLEDLighting.com

CPSIA information can be obtained
at www.ICGtesting.com
Printed in the USA
LVHW021127241118
598135LV00046B/2334